WITNESSING A WOUNDED WORLD

Witnessing a Wounded World

A THEOLOGY OF ECOLOGICAL TRAUMA

Timothy A. Middleton

FORDHAM UNIVERSITY PRESS   NEW YORK 2026

This book is published in an open access edition and released on
University Press Library Open (UPLOpen.com) as part of the "UPLOpen
Climate Change Collection." UPLOpen is an initiative of Paradigm
Publishing and the De Gruyter eBound Foundation.

Copyright © 2026 Fordham University Press

All rights reserved. No part of this publication may be reproduced, stored in
a retrieval system, or transmitted in any form or by any means—electronic,
mechanical, photocopy, recording, or any other—except for brief quotations
in printed reviews, without the prior permission of the publisher.

Fordham University Press has no responsibility for the persistence or accuracy
of URLs for external or third-party Internet websites referred to in this
publication and does not guarantee that any content on such websites is,
or will remain, accurate or appropriate.

Fordham University Press also publishes its books in a variety of electronic
formats. Some content that appears in print may not be available
in electronic books.

Visit us online at www.fordhampress.com.

For EU safety / GPSR concerns: Mare Nostrum Group B.V.,
Mauritskade 21D, 1091 GC Amsterdam, The Netherlands,
gpsr@mare-nostrum.co.uk

Library of Congress Cataloging-in-Publication Data available online at
https://catalog.loc.gov.

Printed in the United States of America

28 27 26   5 4 3 2 1

First edition

To Claire

Contents

Preface   ix

Introduction   1

1   The Traumatized Earth   19

2   Trauma in Ecotheology   41

3   Ecology in Trauma Theology   59

4   The Rupture of Communication: Christ's Witness to a Wounded World   74

5   The Rupture of Flesh: Deep Incarnation and Enfleshed Witnessing   93

6   The Rupture of Time: Witnessing Anthropocene Scars   110

Conclusion   129

ACKNOWLEDGMENTS   143

NOTES   145

BIBLIOGRAPHY   207

INDEX   233

Mark Tansey, *Doubting Thomas*, 1986, oil on canvas, 65 × 54 inches.

Preface

Mark Tansey's *Doubting Thomas* is an arresting image. At first glance, we see a figure in the foreground, a car in the center of the canvas, and a dark fracture that dissects the painting from top to bottom—a fault. The jagged trace across the tarmac looks worryingly fresh; the inward-sloping strata in the background feel unstable and surreal; and the monochrome red amplifies the sense of bleeding and hurt. Something is deeply wrong.

Tansey's painting initially shocks us with its violence, but on closer inspection it begins to prompt a series of deeper reflections. The fault presumably marks the location of a recent earthquake. Read literally, the scene evokes the immediate aftermath of a seismic event. A sudden release of elastic strain energy has ruptured the road and cracked the cliffs behind. Nobody within the immediate field of view looks injured, but we are left to imagine the scale of suffering and tragedy that has been wrought further afield. An indiscriminate violence appears to have torn the land in two, leaving destruction and confusion in its wake. But earthquakes also possess a symbolic resonance. Read metaphorically, this seismic rupture may represent wider nonhuman afflictions, since earthquakes are often interpreted as paradigmatic examples of evil and suffering in the natural world. The connotations of fracture, fragmentation, and transformation make the earthquake a potent symbol of radical schism. It could stand for psychological, social, or cosmic upheaval—or simply function as a reminder of everything that is wrong with the world. The image confronts us with a sign of the planet's general woundedness.

The two human beings in the painting have quite different reactions to the evidence of catastrophe that surrounds them. The woman in the car appears faceless, nameless, and restless, the epitome of an anonymized pursuit of progress. The car itself is sleek and polished, already disappearing off the canvas, apparently impatient to move on. Yet the woman occupies a precarious position. As the car straddles the fissure, she seems oblivious to her own vulnerability. There could be a warning here: technological optimism and human hubris often go hand in hand. By contrast, the man in the foreground appears tentative and concerned. He has stopped to examine the fault, showing it a certain reverence and attention. But he also seems wary as he extends his hand to touch the ground. Crouched in the middle of the highway, he is acutely aware of the danger he might be in, not just from other cars but also from the Earth itself. Intriguingly, the usual gender stereotypes are reversed: it is the woman who has faith in technology, and the man who is in touch with nature.

But the image highlights more than just the natural and human worlds; *Doubting Thomas* is loaded with religious allusions too. The very title of the painting invites us to look again through a theological lens. By labeling the man in the foreground Thomas, the doubting disciple, Tansey sets up a new series of identifications. The body of the Earth becomes the body of Christ; the ruptured fault becomes Christ's wounded side; the scene of recent seismic activity becomes a stark reminder of the crucifixion; and the shockwaves of this earthquake become the traumatic fallout from Christ's death.[1] On closer viewing, one even discovers a cross in the center of the painting, formed by the intersection of the road markings and the fault trace.[2] The whole spectacle is deeply reminiscent of Caravaggio's *The Incredulity of Saint Thomas*. In Caravaggio's rendering we see Thomas, utterly mesmerized and flanked by two astonished disciples, inserting his finger into Christ's wounded flesh. By invoking this art historical parallel, Tansey succeeds in conveying the somatic and affective character of Thomas's encounter with the wounded Earth. Thomas is depicted kneeling at the site of violence, stunned by what he observes, and reaching out for tangible confirmation of his worst fears. In both paintings, it is impossible not to see the wound.

Theological accounts of Doubting Thomas often interpret Thomas as a figure of disbelief who subsequently discovers a faith in Christ. But in the case of Tansey's ruptured Earth, what precisely does Thomas doubt? That the Earth could be so violent? That the Earth could be so vulnerable? That the Earth nonetheless lives on?[3] For Christian trauma theologian Shelly Rambo, it is a mistake to read the story of Doubting Thomas as a triumph of certainty over doubt.[4] The context of Thomas's encounter with Christ's body is one of fear, unfamiliarity, and disorientation.[5] Likewise, Tansey's Thomas by the roadside seems far from certain what he is witnessing. He exists in a state of turmoil and bewilderment.

The painting's theological resonances also contain a decisive twist. In John's gospel, Thomas encounters the *resurrection* body; Tansey is therefore equating Christ's *resurrected* body with the Earth. After whatever traumas Christ/Earth has undergone in the crucifixion/earthquake, this is Christ/Earth in a redeemed state. For Rambo, it is a grave theological mistake to believe that the wounds are only temporary, ultimately erased in some heavenly state of somatic perfection.[6] Rather, Christ's/Earth's wounded body—a body bearing clear signs of trauma—is the only body through which any talk of redemption is possible. In Rambo's words, "some wounds do not go away."[7] The resurrection body is an ambiguous and scarred entity, marked by ongoing faults. Moreover, Thomas's encounter with the wounded resurrection body invites him to examine his own complicity in the wounding.[8] Theologically, Thomas's answer can be parsed in the vocabulary of sin. Ecologically, Thomas might consider the extent of his responsibility for anthropogenic suffering in nature. And we too should ask: have we left permanent wounds on the world?

Tansey's *Doubting Thomas* is a compelling and generative image. It also crystallizes many of the themes that are central to this book. First, if Christ's crucifixion is taken as a paradigmatic example of traumatic experience—as several trauma theologians suggest—then Tansey's painting hints at the possibility that the Earth is traumatized too. This "ecological trauma" need not necessarily refer to literal earthquakes, but it nonetheless expresses something of the Earth-shattering realities that the planet currently faces. In what follows, I read some of these catastrophic realities—such as anthropogenic climate change and biodiversity loss—through the lens of trauma theology. Second, Tansey's image points to a striking parallel between the body of the Earth and the body of Christ. One seems to echo and highlight the other. As I develop a theology of ecological trauma, I suggest how we can envisage God's accompanying presence in the midst of the Earth's traumatic suffering. Finally, Tansey's *Doubting Thomas* foreshadows Rambo's trauma-informed interpretation of the same scene, only with an additional ecological focus. If Rambo is correct, then Doubting Thomas is not a figure who delivers certainties, but someone who returns to a site of unfathomable suffering. He cannot quite formulate what hope looks like in the wake of the trauma he observes, and yet he persists with his presence. Tansey's Thomas is witnessing a wounded world.

WITNESSING A WOUNDED WORLD

Introduction

We are in the midst of a global ecological crisis. In recent years, we have observed a huge range of devastating processes and events: severe storms and hurricanes, intense heatwaves, blazing wildfires, melting ice caps, rising sea levels, habitat destruction, soil degradation, air and water pollution, resource depletion, growing food insecurity, species extinctions, disease outbreaks, and inexorable planetary warming. Many of these phenomena result in such overwhelming suffering—for humans and nonhumans alike—that they are hard to fully comprehend. The whole planetary ecosystem is out of kilter. In the theological arena, ecotheologians have been left wrestling with the sheer scale of these unfolding catastrophes.

At the same time, trauma theorists, and society at large, have become increasingly aware of the incidence of trauma in a growing variety of contexts, from survivors of abuse to collective tragedies. Trauma theologians have therefore had cause to reflect theologically on forms of suffering that are intrusive, recursive, and unresolved. In response to these human traumas, they have offered strategies that can help survivors to cope with ongoing suffering, and trauma-sensitive reinterpretations of theological motifs. Yet an ecological dimension has thus far been largely absent from this work.

In this book, I ask what might be gained by bringing some of the concepts and methodologies employed by trauma theologians to bear on some of the key questions that arise within ecotheology. What kind of traumas are being precipitated by anthropogenic climate change and accelerating biodiversity loss? What would it mean to envisage the Earth itself as traumatized? And how might a Christian theologian respond to these turbulent circumstances?[1] My aim is to demonstrate the need for, and offer one possible example of, *a theology of ecological trauma*.

The rest of this Introduction lays out a few parameters for my analysis. I begin by explaining in greater detail what I mean by "ecological trauma," and I offer a characterization in terms of three diagnostic ruptures—to communication, to flesh, and to time. I then turn to various theological considerations, outlining the trauma-informed methodologies that I seek to employ, and justifying my specific focus on Christology. I end by commenting on the links between ecological trauma and decolonial justice, as well as providing sketches of the chapters to come.

The Concept of Ecological Trauma

It is important to state at the outset what "ecological trauma" denotes. Several definitions are possible depending on who or what is traumatized. It could refer to forms of mental and physical anguish experienced by human beings because of past, or anticipated, ecological destruction.[2] It could refer to well-documented diagnoses of post-traumatic stress disorder (PTSD) in other species such as elephants, chimpanzees, and rodents.[3] It could refer to collective cultural processes of disruption, avoidance, and denial in the face of ecological breakdown.[4] Or it could refer to a trauma that is inherent in ecosystems, landscapes, and natural processes—a trauma that is felt by the Earth itself. Although all four definitions are perfectly legitimate and increasingly relevant in this time of ecological crisis, my primary focus here is on the final possibility: the Earth's own trauma.

Hence, my definition of ecological trauma does not—at least in the first instance—concentrate on human beings. It is widely accepted that disasters like hurricanes, floods, fires, and pandemics can produce traumatic human responses.[5] It is also increasingly recognized that the current ecological crisis is having serious repercussions for human mental health in ways that overlap with the symptoms of trauma.[6] For instance, Panu Pihkala notes how environmental researchers are especially liable to experience psychic numbing, compassion fatigue, and burnout.[7] Ann Kaplan discusses the possibility of climate-induced pre-traumatic stress syndrome (PreTSS), caused by the anticipation and imagination of future ecological threats.[8] And Benjamin White points out that human trauma is not just a symptom of the climate crisis, but also a cause: dissociation from the natural world is a part of what allows us to exploit it.[9] Related terms such as "ecological anxiety," "ecological grief," and "ecological despair" are becoming widespread in popular and academic literatures, and research into these phenomena is of paramount importance.[10] Nevertheless, while my definition of ecological trauma can certainly include ecological events and processes that are traumatizing to human beings, my principal focus here is what happens when we conceive of nonhuman entities—and particularly the planet itself—as the subject of trauma.[11] Given that the Earth is not usually considered to be conscious in any conventional sense, further explanation is required as to how the planet can be understood in this way—and Chapter 1 takes up this task in detail.

Suffice to say, I use the phrase "ecological trauma" to refer to various phenomena and processes within the Earth system that display characteristics analogous to the symptoms of human trauma. Two of the biggest concerns in this regard are climate change and biodiversity loss. The latest report from the Intergovernmental Panel on Climate Change (IPCC) shows that surface temperatures have risen by 1.1°C since the second half of the nineteenth century.[12] It also notes that "the scale of recent changes across the climate system as a whole—and the present state of many aspects of the climate system—are unprecedented over many centuries to many thousands of years."[13] These changes are impacting the frequency and intensity of extreme weather events including heatwaves, heavy precipitation, droughts, and tropical cyclones.[14] What is more, warming of 1.5°C—the preferred limit set by the Paris Agreement—is already on the verge of being exceeded unless greenhouse gas emissions are drastically reduced.[15] Meanwhile, the most recent global assessment of biodiversity concluded that one million species are currently at risk of extinction.[16] We are on the cusp of the planet's sixth great mass extinction event, with rates of extinction tens to hundreds of times higher than they have been for the last ten million years.[17]

Climate change and biodiversity loss can both be considered traumatic because they involve physical wounding, bodily shocks, radical disruption, distorted temporalities, irreversible changes, recurrent suffering, communication breakdowns, and an increasing degree of individual and collective disconnection from normative realities. These symptoms apply differently to different entities depending on the psychic apparatus they are deemed to possess, but the key point is that they constitute forms of ecological suffering that are intrusive, recursive, and unresolved. Moreover, in describing climate change and biodiversity loss as traumatic, we must remember that they each comprise many smaller-scale events and processes that generate specific traumas for specific creatures—and these traumas should be included within the wider definition of ecological trauma. In ascribing ecological trauma to the Earth at a planetary level, it is important not to lose sight of the particularity of the creaturely traumas that are also entailed. Ecological trauma is always already multiple.[18]

It is noteworthy that these examples of ecological trauma both have human origins: climate change and biodiversity loss are driven by human greed, ignorance, and mismanagement. Yet many ecotheologians also deal with ecological suffering that is not directly caused by human beings. Phenomena such as evolutionary suffering and natural disasters could, in theory, be diagnosed as further instances of ecological trauma.[19] However, most trauma theorists insist that trauma does not reside in events themselves, but in an entity's response to those events.[20] Furthermore, some thinkers justify the suffering entailed by evolution or natural disasters on the basis that they contribute to a greater good, whereas trauma scholars typically refrain from describing traumatic suffering as beneficial or purposeful. Trauma is not usually understood as either innate or teleological. Consequently, neither evolution (as a whole) nor natural disasters (in the abstract) are necessarily traumatic phenomena. Individual instances of disease, predation, suffering, or disaster may well become traumatic because of the way they are received and experienced by certain entities, but the natural world is not inherently traumatized.[21] In theological terms, this represents a commitment to the basic goodness of the created world.

Finally, it is important to comment on several other phrases that overlap with ecological trauma. "Ecological trauma" is preferable to "environmental trauma" because the term "environment" risks the impression that the Earth is a passive backdrop to a purely human drama. Ecology, on the other hand, focuses attention on complex relationships: between human and nonhuman, organic and inorganic, and animate and inanimate.[22] "Ecological trauma" is also more helpful than either "planetary trauma" or "climate trauma," which are both more restrictive in scope. "Planetary trauma" invokes trauma on the scale of the Earth but excludes more local instances of ecosystem destruction. Meanwhile, "climate trauma" captures the traumatic suffering of anthropogenic climate change while leaving aside other aspects of the ecological crisis such as biodiversity loss, species extinctions, habitat destruction, unsustainable resource use, damage to nutrient cycles, and so on.[23] "Ecological trauma" is the most versatile of these different terms, and therefore the most appropriate here.

Three Ruptures: Communication, Flesh, and Time

My working hypothesis throughout this book is that it is possible to observe structural similarities between certain symptoms of human trauma and various ecological phenomena. One approach to trauma, proposed and developed by Shelly Rambo and Karen O'Donnell, describes it in terms of three ruptures: to *communication*, to *flesh*, and to *time*.[24] The emphasis on rupture is especially helpful because it connects the medical literature on trauma with literary approaches to trauma theory. It is also a framework that is sufficiently flexible to accommodate a range of traumatized subjects. Here, I explain how these three ruptures encapsulate traumatic experience, first in the case of human trauma, and then in the ecological realm.

The *rupture of communication* forms the first plank of this threefold characterization of trauma. This rupture reflects the fact that many human trauma survivors struggle to express the violence that is inflicted upon them; it seems to be beyond both cognition and language.[25] Medically, the cerebral cortex, which deals with conscious thought, is detached from the limbic and autonomic nervous systems during a traumatic event, enabling a rapid response to danger but preventing a person from processing the event in real time.[26] The original experience is not properly integrated into ordinary memory, leading to a breakdown in that individual's ability to communicate what has occurred. The rupture of communication also points toward the inaccessible nature of trauma that is often emphasized in its more literary treatments. Dori Laub describes this as a "collapse of witnessing."[27] The danger is that any speech, narrative, story, or memory seeks to contain and control trauma; "it understands too much."[28] Yet the persistence and recurrence of trauma mean that it is also impossible to avoid attempts to narrate and remember it. As Cathy Caruth is keen to point out, "the impossibility of a comprehensible story . . . does not necessarily mean the denial of a transmissible truth."[29] Truth may nonetheless be communicated in a disjointed and fragmentary form.

The *rupture of flesh* is arguably the most obvious feature of human trauma. At root, says O'Donnell, trauma is "an experience of injury or invasion of the body."[30] Military casualties or survivors of domestic and sexual abuse, for example, invariably experience a physical rupturing or wounding of flesh. Their flesh is damaged, intruded, or permeated in horrendously painful and undesired ways. But this notion of fleshly rupture is also designed to encapsulate the psychological dimensions of wounding—the mental wounds to the human psyche that can accompany these physiological injuries. The rupture of flesh therefore refers to all the ways in which a survivor's fundamental biology is altered. As Bessel van der Kolk puts it, "the body keeps the score."[31]

Lastly, the *rupture of time* is central to an understanding of human trauma. It includes the fragmentation and failure of traumatic memories from the past; the invasion of the present via flashbacks and nightmares; and an inability to imagine the future.[32] Rambo points out that the temporal distortions of trauma are the antithesis of the familiar statement that "time heals all wounds."[33] Not only do symptoms of trauma return and persist, but distortions of time are themselves part of the wound. In O'Donnell's words, "the normal timelines of cause and effect, of past, present, and future, cease to have meaning."[34] Trauma's rupture of time dismantles the logic of a progressive temporality; time loses its sense of linearity.

Shifting to the ecological realm, the *rupture of communication* figures particularly acutely. In cases of human trauma, the struggle to articulate what has occurred stems directly from the traumatic event itself. Ecologically, the rupture of communication operates slightly differently. Animals might squeal in pain; trees may even be able to release signals via their interlinked root networks; but if we expect communication to take the form of human cognition and language, then the natural world has never had the ability to articulate its suffering. The rupture of communication is not because of any new traumatic events. Instead, we have progressively lost touch with ecological suffering as we have sought to insulate ourselves from the natural world. This rupture is not just a symptom of ecological trauma but also an ongoing problem that will continue to lead to re-traumatization until it is addressed. Laub's "collapse of witnessing" in the ecological realm is our own failure to bear witness to the traumatic wounds of the Earth.

Ecologically, the *rupture of flesh* is again relatively obvious. ("Flesh" here refers to any physical matter.[35]) When an animal on the verge of extinction is shot by poachers or a tree is felled by a commercial logging company, there is a literal rupturing of creaturely flesh. Similarly, deforestation scars, opencast mines, and landfill sites mark the planet's flesh. Even calving glaciers and splintering ice sheets can be imagined as ruptured flesh inasmuch as they betray some of the trauma of a warming climate. These ruptures are wounds in the physical rather than the psychological sense, but, just like traumatic human wounds, they have ramifications that extend beyond their immediate context. What is also noticeable is that, even in the absence of any other forms of communication, the Earth's ruptured flesh may yet convey something of the planet's trauma. Indeed, we might develop van der Kolk's phrase and say that the *Earth*'s "body keeps the score."

Lastly, the *rupture of time* is also evident in the ecological realm. It can be tempting to think of the natural world as a static expanse, or a slowly evolving set of surroundings, but it is sometimes subject to disturbing and distorted temporalities. Climate tipping points provide a dramatic illustration.[36] In the climate system, complex nonlinear feedback loops create lag times such that unwanted symptoms intrude unpredictably. For example, current emissions of greenhouse gases may appear to result in no immediate consequences, but the sudden catastrophic melting of the Greenland ice sheet, which will lead to further warming and a sharp rise in sea level, is already close to occurring due to the cumulative impact of historical emissions. In fact, human impacts on the planet are so pervasive that scientists have been considering whether we have begun a new epoch altogether—the Anthropocene.[37] Our modifications of the Earth's surface are so severe that it appears that they will be preserved in the rock record for millennia to come. As the IPCC notes, "many changes due to past and future greenhouse gas emissions are irreversible for centuries to millennia."[38] The near permanence of our destructive activities is a stark reminder of what Judith Herman calls the "indelible imprint of the traumatic moment."[39]

These three ruptures—to communication, to flesh, and to time—provide a schema for articulating the structural similarities between human and ecological trauma. The second half of this book devotes a separate chapter to each rupture, expanding on the similarities I have just identified, and offering a theological approach to each one in turn.

Approaches in Trauma Theology

Given these outlines of what ecological trauma entails, it is salutary to examine what it might mean to speak of a theology of ecological trauma.

It is important to note straight away that a theology *of* ecological trauma involves more than just a pastoral response *to* ecological trauma.[40] As previous theologies of trauma have made clear, the "lens of trauma" provides an opportunity for reassessing, rethinking, and reconfiguring existing theological ideas.[41] Furthermore, as trauma theologians have progressed in this work, they have drawn attention to the fact that many traumas were already lying latent within Christian scripture and Christian tradition.[42] Trauma is not being imposed on theology, but emerges out of theology. In the same way, a theology of ecological trauma involves not only a pastoral response to ecological devastation, but also a revisiting and reinterpretation of existing doctrines. Indeed, something akin to ecological trauma may already be embedded within Christian thinking.

So, what methodologies do trauma theologians employ that might help ecotheologians to forge a fresh approach to ecological suffering in the context of the current ecological crisis? Comparatively little has been written about methodology within trauma theology even though it has been operating as a distinct subdiscipline of contemporary theology for well over a decade. Part of the issue seems to be that many trauma theologians communicate their approach implicitly rather than explicitly, often beginning from the experiences of survivors and the insights of other disciplines rather than from more abstract frameworks.[43] Nevertheless, there are some common features to draw out.

O'Donnell and Katie Cross position trauma theologies as both constructive and practical.[44] According to O'Donnell, constructive theologies follow four key principles: they recognize that theology is a changeable and evolving discipline; they are willing to draw on resources from beyond Christianity; they result in multiple parallel theologies; and they remain in continuity with the "goods" of the Christian tradition.[45] As such, constructive trauma theologies do not aim to replace previous systematic theologies, but rather seek to build new, open-ended, and imaginative ways of doing theology that engage with contemporary crises and injustices. As Jason Wyman states, "constructive Christian theology acknowledges the constitutive discursive role theologians play in constructing Christianity, rather than supposing that theology attempts to describe an objective deposit of religious truths."[46] Wyman's portrait reinforces the view that constructive theologies are provisional, changeable, and multiple. What they lose in certainty, they stand to gain in relevancy. On occasion, it is undoubtedly the case that constructive trauma theologies are required to undertake "root-and-branch revision" of certain doctrines and read "against the grain of dominant interpretations."[47] But they also attempt to remain faithful to what O'Donnell calls the "goods" of the Christian tradition and can, in the process, unearth fresh readings of doctrine and scripture that may previously have been buried. Meanwhile, practical theologies, write O'Donnell and Cross, "engage critically with the dissonance between theology and lived reality."[48] On this understanding, the worth of a doctrinal claim is measured against the lives of real people. According to Elaine Graham, practical theologies entail: an emphasis on orthopraxis over orthodoxy; a contribution to liberation and justice; and a starting point in experience.[49] Practical theologies therefore tend to look to first-person narratives, and aim to offer pragmatic, if incomplete, solutions. They are driven by the needs of people on the ground. Practical trauma theologies are especially focused in this regard because they aim to foreground the experience of trauma survivors: this is both what drives the desire to theologize and determines acceptable ways to proceed.

Here, my explorations align most closely with the constructive trauma theologies sketched above, where theological ideas are brought to bear on contemporary realities in new ways. More specifically, the following four features, distilled from existing works of trauma theology, each shape my approach in this project.

First, as the overlap with practical theologies emphasizes, trauma theologies tend to focus on the survivor of the suffering. Their needs dictate the theologizing. Others may learn from a trauma theology, but the initial audience is always the survivor. In the case of ecological trauma, this principle poses something of a challenge. According to my definition of ecological trauma, the survivor is the Earth system itself (along with many of its constituent parts). It is therefore the needs of the Earth that establish a framework of permissibility in relation to our theologies. Any theology of ecological trauma must begin from the experience of the suffering Earth and attempt to keep this suffering at the forefront of any theological response.[50]

Second, trauma theologies often prioritize the present moment. They do not typically attempt to explain why traumatic suffering occurs, or how it can be overcome, focusing instead on living with ongoing trauma in the present. It might seem somewhat unconventional for theologians to talk about traumatic suffering without reference to either sin or salvation. Yet the approach here is not to deny the value of discussing origins and solutions, but to seek to redress a theological imbalance. Given that so much has already been written about theodicy, hamartiology, soteriology, and eschatology, trauma theologians often prefer to look for alternative modes of thought. As Deanna Thompson says, the principal question is: "how do I live into this reality that is now my life?"[51] The aim is not to find answers, but to discover ways to cope and to speak theologically in the midst of recurrent and persistent suffering. Likewise, in an ecological context, eschewing both the causes of, and the potential solutions to, ecological suffering seems myopic. Short-sighted presentism is rightly identified as one of the major problems with current ways of thinking and behaving. But the intention is not to discount approaches that try to explain, or look for solutions to, ecological devastation; this sort of research remains vitally important. Rather, the objective is to ask what sort of theological motifs are helpful as we also learn how to live with and accompany this devastation in the here and now.

A third feature of many trauma theologies, which stems from their constructive nature, is a rejection of more systematic theological formulations. Serene Jones is especially candid about her inability to offer up a systematic account. She admits:

> Recognizing this fact [about the painful particularity of trauma] led me to abandon the project of writing a systematic theology of trauma and grace. Rather, what my own writings on trauma continue to seek is a glimpse of grace at work in the interstices of imagination.[52]

The inherently fragmentary nature of trauma makes systematic theologizing nigh on impossible. Logical explanations that are too rigid risk compromising our ability to adequately acknowledge the realities of suffering. This does not mean that theologians should not attempt clear and rational thinking, but it does mean that we should be wary of claims to have completely integrated traumatic experience into a wider explanatory framework. In the ecological context, for instance, this means that a robust doctrine of creation coupled with a systematic account of eschatological redemption for that whole creation is unlikely to be able to address the realities of ecological trauma. Such complete schemes do an injustice to the fractured nature of traumatic existence. Instead, specific theological motifs can offer space for consolation and reflection without pretending to have articulated the inarticulable. There is a place for systematic theologies in other contexts, but, when addressing trauma, theologians must offer a parallel account.

The last point relates to both the focus on the present moment and the desire to resist certain elements of systematic theology—namely, a reconfiguration of soteriology. Rambo glosses a traditional Christian position: "resurrection is the promise of life *after* death that rests on the figure of the resurrected Jesus who triumphs *over* death."[53] But she then seeks to problematize this account. In cases of trauma, she says, the relationship between life and death is more ambiguous.[54] Trauma is not confined to a single event; it persists and returns. Linear models of recovery and healing are not good metaphors for redemption because trauma disrupts such straightforward salvation.[55] Rambo is not dismissing the resurrection, but she is seeking to ask what the resurrection can meaningfully say to survivors of trauma who have no experience of its triumphant aspects. In ecological terms, Rambo's reconfiguration of soteriology points toward some of the dangers of thinking that creation simply needs to be saved; reality is more complex than this. A theology that speaks to life in the midst of death, as opposed to life overcoming death, feels much better suited to contexts of ecological devastation.

There is no singular methodology in trauma theology. But, together, these four features provide a framework for my own construction of a theology of ecological trauma. I aim to focus on the survivors, center the present, resist the temptation to build large-scale explanatory schemes, and remain alert to the need to rethink the trajectory of salvation. As such, I am principally concerned with modes of existing and accompanying—attending to, caring for, and remaining with the wounded.

Christological Motifs

In the second half of this book, I employ a variety of Christological motifs as I seek to articulate a theology of ecological trauma. But it is important to offer a justification for this Christological focus. Why this doctrinal locus as opposed to, say, a theology of creation, or a pneumatological approach? What is it about Christ that is especially helpful in relation to ecological trauma? One reason is simply that Christian theology instinctively turns to Christ; Christ is central to Christian understanding. In general, ecotheologians have been relatively slow to consider the importance of an ecological dimension to

Christology. Yet, as Sallie McFague observes, "since Christology is the heart of Christianity, we must ask whether Christology can be ecological."[56] In fact, there are numerous ways in which Christology contributes to ecotheological discussion via, for instance, Christ's example as an ecological steward, Christ's relationship to the land in the New Testament, cosmic Christology, wisdom Christology, or the implications of the incarnation.[57] Meanwhile, in terms of trauma theology, it is Christ's incarnation in the flesh that is especially pertinent. Christology focuses on the flesh in a way that valorizes matter and brings divinity into communion with the quotidian existence of the physical world. At root, incarnation affirms material being. This is important for a theology of ecological trauma because trauma is a form of suffering that is inherently enfleshed; the rupture of flesh comprises one of its basic characteristics. Furthermore, the theological account of ecological trauma that I develop is built around the concept of witnessing. This idea of witnessing—and cognate notions of accompanying, living with, and showing solidarity toward—is intrinsically incarnational. Hence, it seems likely that a Christological approach to a theology of ecological trauma is a fruitful place to start.

There are also pragmatic reasons for focusing on Christology. There is simply not space here to reflect on multiple theological loci in any detail. Instead, I hope that these Christological reflections can provide a starting point for further explorations in the future. Moreover, as the trauma-informed methodology outlined above particularly emphasizes, it would be inappropriate to attempt a comprehensive and systematic theological treatment of ecological trauma. I also do not envisage providing a singular Christology in response to ecological trauma—and for much the same reason.[58] This is simply not in keeping with the nature of traumatic phenomena. Rather, I aim to offer a series of Christological motifs that bear witness to traumatic ecological devastation.

In offering up these motifs, there is one specific stream of Christological thought that I draw on frequently: the idea of deep incarnation. In brief, deep incarnation theology, first proposed by Niels Gregersen, emphasizes that Christ was not just an isolated human individual, but took on a much deeper connection with the rest of existence in the incarnation.[59] Specifically, Christ's incarnate flesh is understood to be embedded within the *social*, *biological*, and *material* fabric of the universe. *Socially*, Gregersen notes how Christ's body is never presented alone but is always interwoven within cultural networks. This includes his genealogy, his relationship to the landscape, his social interactions with other bodies, his crossing of genetic and cultural boundaries, and his desire to make his body accessible to others.[60] Christ's body is anything other than self-contained; it has ramifications and influences "deep" within the social web of existence. *Biologically*, Gregersen's concern is to emphasize the way in which Christ's flesh is part of the evolutionary process. Christ's flesh is evolved flesh, and so one can imagine extending divine involvement back into evolutionary history. As Gregersen writes, "the incarnation of God in Christ can be understood as . . . an incarnation into the very tissue of biological existence, and system of nature."[61] Perhaps most importantly, deep incarnation theology seeks to emphasize Christ's connection to *materiality*. Gregersen interprets the Word (*Logos*) becoming flesh (*sarx*) in John's gospel as an incarnation into the very substrate of the universe.[62] As Gregersen puts it, "the divine Logos . . . has assumed not merely humanity, but the *whole malleable matrix of materiality*."[63]

Deep incarnation has several helpful aspects for a theology of ecological trauma. As just mentioned, a Christological approach appeals because of Christ's incarnation into the flesh, the very locus of ecological trauma. Gregersen's persistent emphasis on Christ conjoining "all flesh" is therefore particularly helpful.[64] Moreover, in recognizing the social, biological, and material entanglements of Christ's flesh, deep incarnation theology consciously decenters the human.[65] A theology of ecological trauma needs this less anthropocentric, more holistic, understanding of incarnation because it is concerned with the traumatic suffering of the whole planet, and not just the trials and tribulations of humanity. Finally, there is a strong affinity between deep incarnation Christology and supralapsarianism, or "incarnation anyway" theology—the idea that the incarnation would have occurred irrespective of human sin.[66] This view of the incarnation as being about more than just human sinfulness is in concert with the need to respond to traumatic suffering regardless of its origin.

Yet this focus on Christology is not without potential problems; there are reasons to be wary of this turn to Christ. First, I do not mean to imply that Christ is some sort of "answer" to the "problem" of ecological trauma—far from it. Although Christian hope is often understood as being about ultimate redemption in Christ, the trauma-informed methodology that I sketched above encourages us to focus first on present realities. Redemption needs to be rethought, reconsidered, and reconfigured. Otherwise, there is a danger of offering simplistic solutions to complex problems. That is why, in this book, I present Christ not as a harbinger of triumphant healing and cosmic salvation but as a witness who remains in the aftermath of ecological trauma. Christ may well fulfill these other roles too, but in the midst of traumatic ecological suffering it is his appearance as a witness that is most relevant.

Second, an excessive focus on the violence of Christ's cross risks glorifying suffering. As feminist and womanist theologians have long pointed out, if one extols crucifixion as an example, it can encourage passivity in the face of suffering and promote further violence.[67] There is a balance to be struck between Christ's solidarity with the survivors of trauma and any move that recommends or sanctifies suffering.[68] As I seek to develop a model of Christic witnessing, it is particularly important to understand Christ as a witness to suffering that has already occurred or is absolutely unavoidable, rather than as an encouragement to seek suffering or self-sacrifice for its own sake. Suffering may sometimes be inevitable, but it should never be necessary.

Lastly in this series of cautionary notes, there is a danger that invoking a cosmic Christology in response to ecological trauma—as many ecotheologians may be tempted to do—constitutes a colonial act of imposition, not just politically but also epistemologically. Cosmic Christologies emphasize Christ's role throughout creation.[69] They are naturally appealing to ecotheologians because they articulate an innate connection between Christ and the whole of the natural world. But Kwok Pui-lan sounds a note of warning. The image of a cosmic Christ, she says, gained popularity at exactly the time that the leaders of the Roman Empire were seeking justification for their expansive political rule.[70] Then, and on many occasions since, "the claim that Christ is the universal savior contributed to the ideology of colonization of peoples and land."[71] In addition, there is an interpretive imperialism at play in many cosmic Christologies, refusing to allow space for those with other worldviews to pursue parallel interpretations of their experiences according to their own criteria and frameworks. Any Christological exploration of ecological trauma must therefore be wary of appeals to a cosmic Christ. At the very least, such appeals should be bottom-up rather than top-down, generated and employed by those who freely self-identify as Christian.

Ecological Trauma and Decolonial Justice

Recent work in both trauma theory and trauma theology has become increasingly conscious of its Western bias. Academic work on trauma has principally been produced in the United States and Western Europe, even though the experience of trauma has been far more prevalent in other geographical locations.[72] Not only has trauma studies regularly failed to attend to non-Western suffering, and therefore foundered in relation to its own ethical aspirations, but it has also operated on the assumption that its Western construction of the concept of trauma is universally applicable.[73] As Stef Craps puts it, "uncritical cross-cultural application of psychological concepts developed in the West amounts to a form of cultural imperialism."[74] In addition, a focus on event-based models of trauma has tended to impede work on those cumulative and quotidian forms of trauma that are likely to accompany colonial oppression and structural racism. In short, trauma studies needs to decolonize.

Meanwhile, the traumatic impacts of colonialism are legion. As one commentator writes, "colonization is a historical machine of trauma creation and perpetuation that is fully alive in our times."[75] Christina Sharpe specifically employs the metaphor of "the wake" to describe the ongoing aftermath of Atlantic chattel slavery and the persistence of racism in the United States.[76] For Sharpe, "the wake" refers to a whole range of traumatic symptoms—including limited access to healthcare and education, increased likelihood of incarceration and impoverishment, skewed life chances, and even premature death—that all stem from the impacts of slavery.[77] Craps therefore documents concrete proposals for several new diagnoses of trauma, including postcolonial traumatic stress disorder and post-traumatic slavery syndrome.[78]

Trauma theologians have also become increasingly attentive to the twin traumas of racism and colonialism. They have begun to respond to the impacts of "Jim Crow, lynching, and a poverty-to-prison-to-death pipeline."[79] They have started to question Western assumptions about the primacy of the spoken and written word in the formulation of trauma response.[80] And they have indicated how people traumatized by slavery may have developed traumatized theologies that ignore social and political realities in favor of escapist and otherworldly theological formulations.[81] Even Christ's crucifixion can be read as a trauma at the hands of empire.

The recognition of these forms of trauma is especially pertinent here because racist and colonial legacies are also important drivers of climate change and ecological destruction. Massacres of native populations have often been related to ecological devastation.[82] Patterns of colonial behavior are intimately linked to theft of land and therefore have immediate implications for wider ecologies. In the Americas, African slave labor was part of a project to tame and subordinate the natural environment in the name of development and profit.[83] The plantation regime was about razing forests, removing natural vegetation, and destroying natural ecosystems for the sake of agriculture.[84] For James Cone, it is no surprise that racism is deeply interrelated with the desecration of the Earth because, he says, they both stem from the same colonizing logic. As he writes, "the logic that led to slavery and segregation in the Americas, colonization and Apartheid in Africa, and the rule of white supremacy throughout the world is the same one that leads to the exploitation of animals and the ravaging of nature."[85] Likewise, Melanie Harris comments on the "eerie similarities" between the logic of domination at play in the rape and abuse of African women by white oppressors and the logic of domination operative in cases of environmental injustice.[86] Environmental racism also persists today in several forms, such as the location of hazardous waste facilities in predominantly Black neighborhoods, and the exclusion of people of color from environmental policymaking. In what Achille Mbembe terms a "time of planetary entanglement," ecological trauma will always be bound up with the traumas of racism and colonialism.[87] The one form of trauma implicates and imbricates the other. Ecological justice requires decolonial justice too.

This intersection of ecological trauma with decolonial justice forms an important part of the argument at several points in this book. In Chapter 1, the "disenchanted" Western context in which I discuss ecological trauma is itself the product of colonial history. One of the further traumas of colonialism is an additional difficulty in appreciating who or what can be the subject of ecological trauma. In Chapter 3, I document the expansion of trauma theologies and the growing awareness of the interconnected nature of many traumas. The interplay between colonial domination and ecological devastation is a prime example of exactly this sort of interconnection. In Chapter 5, I specifically turn to Cone's theology in *The Cross and the Lynching Tree* to illuminate the nature of Christ's solidarity with instances of traumatic suffering. His Christology is a building block for my own. And, as I just discussed, a cosmic form of Christology carries immense dangers as a response to diagnoses of ecological trauma. Further colonial impositions of Christ's universal relevance are not an appropriate reaction to sufferings caused by a colonial logic of domination. Nevertheless, I remain acutely conscious that there is further work to be done to develop a theology of ecological trauma within an explicitly decolonial framework.

The Book in Outline

This book is divided into two parts. The first half (Chapters 1–3) explains how and why we might envisage the Earth as traumatized and makes the case for a theological consideration of ecological trauma. The second half (Chapters 4–6) then demonstrates how a Christian theologian might respond, providing one possible theology of ecological trauma.

In Chapter 1—"The Traumatized Earth"—I ask whether it is necessary to ascribe subjectivity to the Earth to be able to describe it as traumatized. In conversation with Joanna Leidenhag, James Lovelock, Theodore Roszak, Graham Harvey, and others, I review some of the literature on the world soul, panpsychism, Gaia theory, ecopsychology, and Christian animism, but I conclude that, at least in "disenchanted" Western cultures, it is advisable to remain agnostic about the existence of a planetary consciousness. Instead, I propose that ecological trauma should be understood as a vitally important anthropomorphism that helps humanity to better relate to the ecological suffering that is occurring. This is not just poetic fancy, but a psychological strategy for negotiating real relationships. I also ground this stance in several scriptural and theological examples of anthropomorphism, including the mourning of the land in the Hebrew Bible, and the cry of the Earth in Pope Francis's *Laudato Si'*.

In Chapter 2—"Trauma in Ecotheology"—I indicate what the notion of trauma can add to existing ecotheological discussions about suffering and woundedness. I observe how the language of trauma helps to register the seemingly inexpressible severity and urgency of the current ecological crisis in a way that has not been possible for many existing theological and ethical frameworks. I also explain how trauma theologies provide a fresh approach to ecological suffering, in contrast to both environmental ethics and natural theodicy, by focusing on learning to live with continuing traumas in the present rather than categorizing suffering based on its causal origin or potential solution. I end the chapter by reviewing five brief engagements with trauma from within the existing ecotheological literature—from Pamela McCarroll, Catherine Keller, Danielle Tumminio Hansen, Mark Wallace, and Matthew Eaton—to tease out how a theology of ecological trauma might proceed.

Chapter 3—"Ecology in Trauma Theology"—also aims to illustrate the need for a theology of ecological trauma, but approaches the question from the opposite direction. Initially, I highlight the fact that Serene Jones already draws attention to the relevance of trauma theology for the ecological realm. As a relatively young subdiscipline, trauma theology continues to expand and evolve, and scholars are still discovering its applicability in new contexts. Moreover, as Shelly Rambo's case study of Hurricane Katrina makes clear, trauma processes often integrate the "human" and "natural" worlds. There are also several ecological elements already embedded in the trauma theology literature, from the nonhuman laments of the Hebrew Bible to the ecological events accompanying the crucifixion.

In the second half of the book, I advance one possible theology of ecological trauma in a Christological key, highlighting how different Christological motifs—such as incarnation, crucifixion, and recapitulation—can help Christian theologians respond to the traumatic ruptures of communication, flesh, and time. Following existing work in trauma theology, I center my discussion on the practice of witnessing. Over the course of Chapters 4, 5, and 6, I propose and develop a model of Christic witnessing as a viable response to ecological trauma.

In Chapter 4—"The Rupture of Communication: Christ's Witness to a Wounded World"—I focus on the traumatic breakdown of communication.[88]

This "collapse of witnessing," as Dori Laub describes it, suggests that we must cultivate a practice of bearing witness in the aftermath of ecological trauma. Following Rambo, this witness to trauma is not a confident proclamation or imitation of the situation, but a humbler and more difficult resolve to "remain" with the suffering. I argue that Christ can fulfill the role of witnessing ecological trauma, in his life and ministry, as well as in his crucifixion and death. I also situate this model of Christic witnessing in relation to other theological proposals about the cruciform creation, divine companionship, co-suffering, Christ as microcosm, and God as traumatized, drawing on the work of scholars such as Holmes Rolston, Ruth Page, and Christopher Southgate. Crucially, Christ's witness is not just a passive presence, but an active stance that creates something new: a record and acknowledgment of traumatic suffering that would otherwise be lost or occluded.

Chapter 5—"The Rupture of Flesh: Deep Incarnation and Enfleshed Witnessing"—turns to the traumatic tearing of physical tissue. Flesh is the locus of trauma, which is why Christ's assumption of flesh at the incarnation is so significant. Deep incarnation theologians—including Niels Gregersen, Elizabeth Johnson, and Denis Edwards—specifically argue that Christ's assumption of flesh can be interpreted as an assumption of materiality itself. I therefore propose that Christ can serve as an enfleshed witness to ecological trauma because he shares in that which he witnesses. I also position this Christic witnessing as a middle way between Christ as a distant representative of creation's pain, and Christ as incarnate in all that exists. I envision Christ to be in the process of incorporating the wounds of the world. What Christ bears in his flesh is a somatic and haptic communication of the trauma being wrought on the flesh of the Earth.

Finally, Chapter 6—"The Rupture of Time: Witnessing Anthropocene Scars"—addresses the temporal traumas associated with this proposed geological epoch.[89] I begin by illustrating how conceptions of linear time are ruptured by both the recurrence and the permanence of various ecological phenomena, not least in the enduring imprint of the Anthropocene within the rock record. A focus on helical time, as demonstrated by recapitulative Christologies, can help us to come to terms with the recurrence of traumatic phenomena in the ecological realm. Meanwhile, the deep time perspective of the deep incarnation theologians speaks to the permanence of ecological trauma. Christ's permanently wounded flesh is a durable witness to ecological trauma that persists deep into the planetary future.

In general, theologies of ecological trauma typically focus on coping rather than solving. In climatic terms, this translates into a focus on adaptation as well as mitigation. But it is important to stress that this is not a fatalistic vision.

A radical hope is possible, through imagination, and through a witness's commitment to remain at the site of trauma. No specific ethical actions are prescribed, and yet the very act of witnessing evokes a sense of responsibility. This is modeled for Christians by Christ, and is echoed in both liturgical repetitions and eucharistic celebrations. It is also a task that all people are called to—as scholars, scientists, activists, and human beings. We should bear witness to a wounded world.

1
The Traumatized Earth

Therefore the land mourns, and all who live in it languish; together with the wild animals and the birds of the air, even the fish of the sea are perishing.[1]

—HOSEA 4:3

It sounds eerily reminiscent of a mass extinction event. Sometimes described as a "prophet of doom," Hosea presents us with the deaths of wild animals, birds, and fish. The suffering extends to the supposedly inanimate world too: even the land itself is mourning, imbricated in this cataclysm. Bound up in relationship with these creatures, the land offers a lament on their behalf as they perish. The scene is one of devastation and distress.

But would it be too much to suggest that what Hosea offers in this epigraph is an account of ecological trauma?[2] Given that trauma theory is deeply rooted in human psychology, is it possible to view the Earth itself as the subject of traumatic experience in the way that Hosea seems to imply? In short, how is ecological trauma to be conceived?

In this chapter, I attempt to steer a course between competing answers to these questions. On the one hand, plenty of people in the modern West would explain Hosea's account, and other similar scriptural examples, as mere poetic license. The land cannot mourn, or experience trauma, in any literal sense because it is not capable of conscious experience. On the other hand, many philosophies and worldviews—positions that can loosely be described as "animist" in their orientation—would feel more comfortable taking Hosea's mourning land at face value. Yet this purported antithesis between modernist and animist ontologies is something of a false dichotomy. For those who are primarily worried about ecological concerns, rational beliefs are much less important than relational behaviors; it is not, in the first instance, what we think that matters, but what we do. When considering the traumatic devastation that Hosea describes, the core of the issue is not whether the land is believed to be capable of certain experiences, but rather whether it is related to as such. My suggestion is that ecological trauma is best conceived as part of a relational anthropomorphism: a vital tool that helps us to imagine, and improve our relationships with, nonhuman others.

Biblical Anthropomorphism

Hosea's mourning is far from an isolated example. The theme of the land's lament recurs numerous times in the Hebrew Bible, and anthropomorphic descriptions of nature are found throughout the biblical canon. Given that the mourning land motif is arguably the closest biblical analogue for ecological trauma, it is instructive to consider some further examples:

> I looked on the earth, and lo, it was waste and void. . . . Because of this the earth shall mourn, and the heavens above grow black.[3]

> How long will the land mourn, and the grass of every field wither? For the wickedness of those who live in it the animals and the birds are swept away.[4]

> The fields are devastated, the ground mourns; for the grain is destroyed, the wine dries up, the oil fails. . . . Even the wild animals cry to you because the watercourses are dried up.[5]

Here we have wasting, withering, devastation, and destruction—and the Earth mourns these potential symptoms of ecological trauma. The prophets describe failed harvests, droughts, and increasing desertification.[6] Invariably, they also link these disasters to human sin and divine judgment. Yet there is some disagreement in the literature about the precise cause of these calamities.

Katherine Hayes asserts that the prophets are primarily referring to the way that social abuse and disorder in the human realm harm the Earth indirectly via YHWH's anger.[7] Hosea, for example, states that it is human beings' misdemeanors (swearing, lying, killing, stealing, and adultery) that the land mourns, even though none of these actions has a direct impact on the land itself.[8] By contrast, Christoph Uehlinger makes the opposite case, arguing that political, military, and social conflicts damaged the ecological sphere directly in the time of the prophets.[9] Social sin—in the form of the over-use of soil, agricultural mismanagement, or war—is deemed to be the proximate cause of ecological disaster without needing to go via the anger of the deity. This reading is much closer to the situation today, where human actions such as effluent discharge, soil pollution, clear-cutting, mining, and monoculture directly abuse the land. Uehlinger's explanation is arguably preferable to Hayes's version because it does not implicate YHWH in the perpetration of ecological trauma. But, either way, planetary mourning ultimately stems from human wrongdoing.

What is more important, though, is the recognition that there are multiple scriptural passages in which the Earth appears traumatized. While these passages do not explicitly use the vocabulary of trauma, the symptoms are in keeping with what one would expect if the Earth were experiencing trauma: emptiness, annihilation, and a residual moaning complaint. It is especially noteworthy that the Earth is attributed the necessary subjectivity and agency to undergo and articulate this suffering. The Earth is granted its own emotional life.

Instances of the mourning Earth are dotted throughout the prophetic literature, but the most common scriptural anthropomorphism is actually the ascription of praise to the nonhuman creation. Terence Fretheim offers no fewer than fifty-two separate examples of nature praising God, primarily from Psalms and Isaiah.[10] The day speaks; the night declares; the valleys shout for joy; the mountains praise; the floods roar; the Earth rejoices; the trees sing; the wilderness is glad; the desert lifts up its voice; the trees of the field clap their hands; and all flesh worships.[11] In Psalm 148 in particular, the sun, moon, stars, high heavens, heavenly waters, sea monsters, deeps, fire, hail, snow, frost, wind, mountains, hills, fruit trees, cedars, beasts, cattle, creepers, and birds are all exhorted to praise the Lord.[12] All manner of nonhuman entities are given humanlike capabilities.

New Testament examples of anthropomorphized nature are far fewer in number. In the gospels, as Jesus enters Jerusalem, the Pharisees ask him to stop his disciples from calling out and he replies, "I tell you, if these [people] were silent, the stones would shout out."[13] The envisaged lithic clamor is a cry of both praise and protest.[14] Meanwhile, in Romans, Paul states that "we know that the whole creation has been groaning in labor pains until now."[15] Again, it is the nonhuman creation that is the subject of all the relevant verbs in this passage: it waits with longing, it is subjected to futility, it will be set free, it groans, and it has been in travail.[16] In Sigve Tonstad's analysis, "non-human creation is subject, not object, speaking as a sentient being that is capable of experiencing suffering and expressing hope."[17] In actual fact, the groaning of creation in Romans is not ideal as an analogue for ecological trauma because it offers a teleological account of nonhuman suffering: like a woman giving birth to a child, the creation is only groaning because it is about to be set free and reveal the glory of God, whereas ecological trauma describes suffering that is both unnecessary and persistent.[18] But the central point is that the whole of creation is attributed the ability to undergo suffering and to express its distress.

So, how are we to interpret these biblical anthropomorphisms? Should we be building a case for a robust biblical animism, or are such passages to be downplayed as mere poetic fancy?[19] And do these scriptural vignettes provide a sufficient basis for thinking of the Earth as traumatized?

A helpful hermeneutical approach is offered by the Earth Bible Project. Enshrined as one of their six ecojustice principles is the statement that "Earth is a subject capable of raising its voice in celebration and against injustice."[20] They elaborate:

> We use the term "voice" as shorthand for the diverse ways in which Earth and Earth community may communicate. . . . By the voice of Earth we mean the many languages of Earth—be they gesture, sign, image or sound—that send a message, whether to humans, to other members of the Earth community, or to God.[21]

For the Earth Bible Project, the Earth has a voice inasmuch as it issues communicative signs, but this group of scholars is also clear that the Earth does not vocalize its concerns in the same manner as human beings. Likewise, they write that "we can speak of Earth as a subject without assuming that Earth has the same consciousness or character as humans."[22] They are clear that any Earth consciousness is utterly different from human consciousness and so should not be governed by human norms. The best we can do as human beings is to relate to the Earth's consciousness and the Earth's voice via anthropomorphic metaphor:

> Anthropomorphic language is the primary mode by which humans communicate to others to whom a significant level of independence, equality and similarity is attributed. The term "voice" is our way of recognizing Earth as communicating as an equal but different "thou". . . . The metaphor of voice . . . helps us—as humans—to grasp and appreciate the reality of communication with a "thou" other than ourselves.[23]

In other words, metaphors about the Earth's voice are a concession to our limited and peculiarly human modes of interaction and communication. Since we cannot hear the parched ground during an extreme heatwave in any literal sense, we must resort to a metaphorical description of the land's mourning.

Ultimately, we cannot second-guess the intentions or belief systems of the biblical authors; they may or may not have held to animist ontologies. But these examples do demonstrate—at the very least—that there is biblical precedent for the use of anthropomorphic metaphor as an approach to our relationship with the natural world. Regardless of whether nature was *believed* to be personal, it was certainly *perceived* and *related to* as such.

The Cry of the Earth

The existence of biblical anthropomorphism provides an initial basis for taking seriously the claim that the Earth itself can be the subject of ecological trauma. But it is important to emphasize that this way of speaking about the Earth is not just confined to biblical sources; it has also been widely adopted in recent theological work. The "cry of the Earth"—as opposed to its mourning or lamenting—only occurs once in the Bible in the book of Job, but it is a motif that has become widespread in contemporary theological writing, especially since the 2015 publication of Pope Francis's encyclical letter *Laudato Si'*.[24]

In an echo of Saint Francis's *Canticle of the Creatures*, Pope Francis writes about "brother sun, sister moon, brother river and mother earth."[25] He says that our common home, Earth, is "like a sister with whom we share our life and a beautiful mother who opens her arms to embrace us."[26] Ecofeminists may legitimately be concerned about the gendered language that is invoked here, but the familial anthropomorphisms are nonetheless profound for the light they shed on how Francis thinks we should relate to the Earth. Part of what Francis probably intends is a gesture of reconciliation toward Indigenous communities for whom Pachamama, or Mother Earth, is a central component of their spirituality. Yet the ecological crisis is also wreaking serious harm on sister Earth. As Francis puts it, this "sister now cries out to us because of the harm we have inflicted on her."[27] Mistreatment of our common home has "caused sister earth, along with all the abandoned of our world, to cry out, pleading that we take another course."[28]

The idea that the Earth can "cry out" is at the core of Francis's proposal in *Laudato Si'*, and it points in two important directions. First, the cry of the Earth is inextricably tied up with the cry of the poor. As Francis says, "a true ecological approach always becomes a social approach . . . so as to hear both the cry of the earth and the cry of the poor."[29] The crucial point is the indivisibility of ecological and social liberation. All too often, desecration of the land goes hand in hand with increasing oppression of those laborers and workers who are most marginalized in society. The cry of the Earth is not separate from, or other than, the lamentations of the poor and their plight. In fact, we may even hear the cry of the poor *as* the cry of the Earth. Since the ecological crisis exacerbates extreme poverty, the two situations are communicated in a single exclamation. One way of ascertaining what is traumatic for the Earth is by listening to what is said by those human beings who are most oppressed. The second important implication of Francis's crying Earth is the voice that it gives to the nonhuman realm. Much like the biblical anthropomorphisms described above, if the Earth in some sense cries out, then it also appears that it is capable of undergoing suffering—and trauma.

In a short article about *Laudato Si'*, Bruno Latour pushes this interpretation further. He notes how, in the encyclical, "our Sister, Mother Earth," is granted "a power to act, a capacity to suffer, to be hurt, to groan, which this time becomes literal rather than metaphoric."[30] Latour is convinced that Francis intends something quite radical here, something that goes beyond mere metaphor and acknowledges that the Earth is capable of suffering in a literal sense. In fact, Latour reads Francis (and the whole encyclical) as "channeling this immense cry" of the Earth.[31] Yet Francis himself inhabits ambiguous territory at this juncture. He is clear that he does not want to "put all living beings on the same level" or "imply a divinization of the earth," and so it remains unclear what sort of ontological weight should be attributed to his language about the cry of the Earth.[32]

Nevertheless, in the context of this discussion, definitive ontological statements about the attributes and abilities of the Earth are ultimately less important than the way in which a certain use of language affects how humanity relates to the Earth. Francis's ultimate hope, as expressed in the prayer at the end of the encyclical, is that "we may protect the world and not prey on it, that we may sow beauty, not pollution and destruction."[33] Francis is more interested in exhorting certain relationships and practices than in prescribing any specific metaphysical scheme, and this is well served by his anthropomorphic language. Beliefs are held to be secondary to behaviors. This is also very much in consonance with the liberation theology and the "theology of the people" that inform Francis's thinking. Orthopraxis precedes orthodoxy.

Moreover, this focus on relationship and praxis appears to be what Saint Francis had in mind in his *Canticle of the Creatures*. For example, Roger Sorrell, in his reading of the *Canticle*, claims that Saint Francis would certainly not subscribe to any pantheist or panpsychist ontology. Instead, what is significant is the cultivated mode of relationship. According to Sorrell, Saint Francis's anthropomorphic language serves to "link humankind with creatures in a positive emotional manner, aiding people to identify with them and feel their kinship with them."[34] What is important here is emotional connection, identification, kinship, and affective relationality. The purpose of this language is to create specific bonds with other creatures and other entities.

Is the Earth Conscious?

So, where does this survey of biblical and theological anthropomorphism leave us? For Sorrell, and others, there is a reluctance to take anthropomorphic language as an affirmation of anything ontological, regardless of whether it appears in Christian scripture or Christian tradition. In this case, it is only possible to talk about the traumatized Earth in metaphorical terms. But for Latour, and others, there is something literal about the cry of the Earth. If the Earth is deemed a subject in a similar manner to human subjects, then it is reasonable to suppose the Earth is also susceptible to traumatic suffering.

I will return to the importance of anthropomorphic metaphor below, but first it is worth pursuing whether a Christian theologian can say anything ontological about the consciousness of the Earth. A wide variety of different spiritualities, traditions, and philosophies allow for some sort of soul or spirit in nature—and there is not space to do any of them justice here. But, in what follows, I offer some abbreviated sketches to illustrate how a range of different positions could offer Christian theology a metaphysical basis for speaking of the Earth's trauma. These various viewpoints—namely, the notion of a world soul, Gaia theory, panpsychism, ecopsychology, and animism—all share an important link: they all posit some degree of conscious behavior beyond the human, which in turn allows for the identification of possible sites of subjective experience that can undergo ecological trauma.[35]

The World Soul

The idea of a world soul (*anima mundi*) that inhabits the world in the same way that human souls inhabit human bodies goes back at least as far as Plato's *Timaeus*. Plato himself describes the world as "a creature with life, soul, and understanding."[36] The purpose of the world soul is as a kind of rational principle, a governing intelligence that mediates between the divine perfection of the forms and the turbulent imperfections of the cosmos. It is also a notion that has proved popular with numerous Christian thinkers, from Origen's Late Antique thought, and the Neoplatonism of Boethius and Eriugena, via the Renaissance writing of Ficino and Bruno, to a Romantic revival by the likes of Schelling.[37]

At first blush, a world soul seems like the ideal locus for the subject of ecological trauma. If possession of a soul—however that is defined—includes the psychological components that are necessary for the experience of trauma, then the world soul might feel the traumatic symptoms of Earth system processes in an analogous fashion to the way that the human soul undergoes human trauma. This simple analogy builds on the oft-cited structural similarities between the human microcosm and the worldly macrocosm: properties of the world at large, such as the ability to experience trauma, may be inferred from the capacities of human beings. But there is also a difficulty in assuming such a straightforward attribution of trauma to the world soul. The Platonic *anima mundi* is fundamentally rational, bringing order out of disorder. It is therefore unclear whether such an organizing principle could be subject to the fragmentation and chaos entailed by trauma. The very definition of the Platonic world soul, built upon a sense of cosmic harmony, seems prearranged to exclude those processes and phenomena that are viewed as traumatic.

Within Christian theology, Plato's world soul has often been shunned on the grounds of paganism. At the same time, the *anima mundi* has also been equated with both the Holy Spirit and the Johannine *Logos*.[38] For example, Athanasius is thought to have borrowed the idea that the *Logos* is the soul of the universe from the Stoics.[39] The world soul is taken to refer to the way in which divine reason indwells the world. One key difference between the Christian understanding of *Logos* and the Stoic view is that for Christians the *Logos* is also deeply personal, as manifest most explicitly in Christ's incarnation. Indeed, if credence is given to this identification between the *Logos* and the *anima mundi*, then it is possible to suggest that Christ is the subject of ecological trauma. God, in the second person of the Trinity, as *Logos*, is the soul of the universe and therefore undergoes the traumatic suffering of the Earth. God, in Christ, experiences ecological trauma.[40] Yet there is a similar difficulty here to that encountered with the Platonic world soul. Christian understandings of the *Logos* often retain the view that it is associated with rationality and order, making it difficult to conceive of the *Logos* as the subject of disordered and traumatic experiences.

Beyond Christianity, Plato's world soul has permeated several other traditions and worldviews, including branches of paganism and animism.[41] If the *anima mundi* is able to shed its connotations of rational order and control, then it is certainly a viable candidate for a robust ontological account of the subject at the heart of ecological trauma. But, for Christian theologians, the association between the *Logos* and divine order makes it challenging to conceive of the world soul as a seat of traumatic experience.

Panpsychism

Panpsychism holds that mind is fundamental throughout the physical universe; all things (*pan*) have mentality (*psyche*). Panpsychist philosophies are often linked to Plato's world soul, although they are also found in several other thinkers and traditions. In recent decades, panpsychism has experienced something of a philosophical revival as a potential answer to the hard problem of consciousness.[42] If the building blocks of mind are common to all matter, then this could explain the origin of human consciousness. Furthermore, if evolution is to operate without any unexplained jumps, then the raw elements of consciousness must have been present from the outset.[43]

Panpsychism is not a position that has been widely considered within mainstream Christian theology, but Joanna Leidenhag makes the case that panpsychism is perfectly compatible with many orthodox Christian beliefs about creation.[44] Moreover, Leidenhag proposes that a panpsychist hermeneutic can help to explain the many scriptural examples of creation's agency and activity (as described above). Biblical examples of the voices of nature almost appear to presuppose a panpsychist metaphysic.[45] This "democratization of mind" is also appealing because if subjectivity is widespread throughout the universe, then this seems to ensure that intrinsic value is a similarly ubiquitous feature of reality.[46] If all parts of creation contain elements of consciousness, then all parts of creation are due moral concern. In addition, Leidenhag suggests that panpsychism provides a unique ontological space for the indwelling of the Holy Spirit throughout the created order.[47] Divine action is enabled by the presence of the Holy Spirit within the mental building blocks of creation.

As with the world soul, panpsychist philosophies seem to provide a neat metaphysical scheme for substantiating the attribution of trauma to the Earth. If mentality is found throughout creation, then it is relatively straightforward to claim that the mind of the Earth can be subject to traumatic experience. Again, however, there are certain difficulties that need to be overcome. Most panpsychists do not typically claim that rocks and telephones are conscious in the sense of having a unified consciousness with functional abilities, but rather that these entities contain nonunified elements of mentality.[48] This leads to what is known as the combination problem. The elements of mind are present in all matter in a dispersed form, but the factors that determine how these elements are arranged to produce macro-level consciousness are left unexplained. One solution, known as cosmopsychism, proposes that the whole cosmos is the fundamental unit of reality that possesses conscious experience, although this position then has the opposite problem of explaining how a universal consciousness is divided up into individually conscious subjects. Either way, most panpsychists do not usually imagine that the Earth itself possesses a unified consciousness. Depending on one's understanding of trauma, this lack of a stable subject at the planetary scale may or may not be a stumbling block for an ontological grounding of ecological trauma. If a unified and preexistent subject is taken to be a prerequisite for the attribution of trauma, then panpsychism is less helpful than it first appears for the metaphysics of planetary trauma. But it is also conceivable that panpsychism enables ecological trauma to be ascribed to a fragmentary and distributed Earth consciousness, or that the Earth's trauma is felt in one part of a cosmopsychic universe. Panpsychist philosophies therefore remain a plausible possibility for theorizing ecological trauma.

Gaia Theory

Gaia theory, as formulated and defended by James Lovelock and Lynn Margulis, asserts that living organisms and their surroundings function as a single self-regulating system that works to maintain conditions that are favorable for life on Earth. Lovelock's early work investigated how organic life interacts with inorganic systems to produce stable concentrations of atmospheric gases such as oxygen, methane, and carbon dioxide in proportions that are conducive to life. Margulis subsequently showed how microbes are particularly important contributors to the Earth's self-regulation. Lovelock initially spoke in terms of an Earth feedback hypothesis, but he was later persuaded by his friend, the novelist William Golding, to name his theory after the Greek goddess Gaia.

The basic science involved has been debated, but it has also achieved widespread acceptance among contemporary Earth system scientists. Meanwhile, Lovelock's invocation of Gaia has given rise to a series of more metaphysical and spiritual claims. For example, Lovelock himself asserts that "the entire range of living matter on Earth . . . could be regarded as constituting a single living entity . . . endowed with faculties and powers far beyond those of its constituent parts."[49] Here, Lovelock seems willing to talk about Gaia as alive and intelligent, if not actually conscious.[50] Others have taken Gaia theory further. For instance, Stephan Harding concludes from Gaia that "matter is sentient to its deepest roots," and that "everything is capable of experience, including . . . the Earth itself."[51]

If the interpretations of writers like Harding are taken at face value, it is easy to see how Gaia theory might underwrite a concept of Earth trauma. For example, Louis Heyse-Moore proposes that "disturbances in the systemic homeostasis of Gaia," such as the disruption of feedback loops in the climate system, constitute what he calls Earth Systems Stress Trauma (ESST).[52] If Gaia is accepted as alive and sentient, then it follows that Gaia is susceptible to the planetary equivalent of psychological trauma, including deforestation, atmospheric pollution, overfishing, and so on. Furthermore, phenomena such as extreme weather events, rising sea levels, and shrinking ice caps can be interpreted as not just traumatic afflictions, but symptoms of a wider trauma process as the Earth attempts to restore its homeostatic balance.[53] The "revenge of Gaia," as Lovelock titles one of his books, could be read as a traumatic fight response on the part of the Earth.

Yet Gaia theory has also come in for significant criticism. As one particularly pointed comment puts it: "Gaia is just an evil religion."[54] Much has been written about how rejections of Gaia theory (often unwittingly) buy into a series of colonial and misogynistic tropes.[55] But this did not stop Lovelock from feeling that, in order to persuade other scientists to take him seriously, he must consistently reemphasize that Gaia is "just" a metaphor. In the preface to his first book on Gaia, for example, he says that the planet is alive and sentient in the same way that a gene is "selfish," or a ship is called "she."[56] Elsewhere, he is forthright when he says that he is "not thinking in an animistic way of a planet with sentience."[57] Lovelock places Gaia theory unambiguously in the realm of anthropomorphic metaphor, but he notes (wryly) how science is often "riddled with metaphor."[58]

At the same time, Lovelock is reluctant to relinquish the metaphor of Gaia altogether in favor of the more scientifically orthodox terminology of Earth system science or biogeochemistry because, he says, "we need the poetry and emotion that moves us."[59] Lovelock insists that the metaphor of Gaia bolsters public understanding and practical environmentalism precisely because it encourages us to treat the Earth as active subject rather than passive object.[60] "Metaphors are more than ever needed," says Lovelock, "for a widespread comprehension of the . . . lethal dangers that lie ahead."[61] As Mary Midgley puts it, the idea of Gaia can be a "powerful tool," a "guiding myth," and an "imaginative vision."[62]

Depending on how one chooses to interpret the significance of Gaia theory it may, or may not, provide an ontological basis for speaking of Earth trauma. Yet, even if Gaia is taken to be "just" a metaphor, there is good reason to think that Lovelock is right about the power of such metaphors to galvanize change, not least because Gaia propelled his own ideas into the popular imagination in a way that he had never dreamed would be possible. In this vein, Gaia theory may still be a "powerful tool" for envisioning the traumatized Earth.

Ecopsychology

The historian Theodore Roszak is often credited with inventing the term "ecopsychology"—in his book *The Voice of the Earth*—to describe the interdisciplinary study of an emotional connection between humans and nature.[63] As Roszak puts it, "the needs of the planet are the needs of the person" and "the rights of the person are the rights of the planet."[64] The centrality of this connection between our inner mental states and the outer exterior world is such that "our humanity is incomplete until we have established our kinship or social relations with the larger natural world."[65] In elaborating this connection, there is even a sense that humanity is simply the self-conscious part of a much bigger system. "The planet thinks through us," says Roszak.[66] Or, in Joanna Macy's words, "we are our world knowing itself."[67]

There is nothing explicitly theological, or even necessarily religious, about ecopsychology. But, as a growing discipline, it also seems especially germane to the idea that the world as a whole is, in some literal sense, traumatized. For ecopsychologists, it is not so much that there is a discrete and separate world soul or planetary organism that is the subject of ecological trauma, but that all consciousness is interconnected. This large-scale interconnection weaves humanity and nature together, providing a distributed consciousness that could be the seat of traumatic experience.[68] Everything is, to some extent, a part of the trauma process. On this understanding, the breakdown of communication that characterizes traumatic experience is especially stark. Western science's increasing objectification of the natural world not only ruptures any connection between humanity and nature, but also leads to increasing fragmentation of the distributed subject that is the very seat of traumatic experience. It is more than deeply ironic that this particular symptom of trauma—the dissolution of the subject—makes it harder to even identify ecological trauma because of an apparent lack of a suitable subject. Part of the goal of ecopsychology is to bring healing to this growing emotional rift between humanity and nature.[69]

By and large, ecopsychology has not had much interaction with the Christian tradition. But the fact that ecopsychologists tend to see human beings as the conscious part of a much larger system might appeal to those Christian theologians who are worried that talk of Earth trauma requires a completely flat ontology.

Animism

Graham Harvey defines animists as "people who recognise that the world is full of persons, only some of whom are human, and that life is always lived in relationship with others."[70] If the world is full of persons, then there is the possibility that these other persons also experience traumatic suffering. As with the other brief sketches above, space does not permit serious engagement with the existing literature on animism—and I am especially conscious that many Indigenous traditions contain a wealth of expertise that does not feature here. However, my (much narrower) concern is to highlight how a Christian theologian might plausibly appeal to animism to ground an understanding of ecological trauma. Indeed, we have already seen how the anthropomorphic language of scripture may prompt consideration of a biblical animism.

Mark Wallace specifically entertains what he calls Christian animism, and notes how it enables an appreciation of the Earth's trauma.[71] Wallace explains how our objectification of the Earth's "resources" means that "Earth is no longer [deemed] a 'living being' or 'feeling organism' who can undergo traumatic suffering."[72] We have, he says, "effectively rendered our living planet numb and silent."[73] To counter these de-animating and objectifying tendencies, Wallace aims to "*re-imagine* Earth as an animate being, a living soul, who feels joy and suffers sorrow and loss just as we do."[74] His rationale for revisiting the Earth's animacy depends on both the findings of Earth system science and the many scriptural examples (cited above) of the Earth's apparent agency and subjectivity.[75] For Wallace, it is clear that the Earth is to be seen as both animate and capable of experiencing trauma.

Wallace's animism is specifically Christian because of the striking role he gives to the Holy Spirit. For Wallace, it is the Spirit who is the "soul" of the Earth. As he writes, "everything God made is a bearer of the Holy Spirit."[76] The Earth is deemed not just personal, but properly divine. Yet he eschews pantheism, arguing that God's presence in the world does not necessitate a complete elision of the difference between God and the world.[77] Nevertheless, on Wallace's account, it is the Spirit that becomes the natural subject of ecological trauma: "God as Spirit undergoes deprivation and trauma through the stripping away of Earth's bounty."[78] This comes with serious theological risks, since it "may result in permanent trauma to the divine itself."[79] But if we lay aside any of the hesitations that Christian theologians often have about the passibility, or otherwise, of the Holy Spirit, Wallace's insistence on the presence of the Spirit in nature provides sufficient personality for a diagnosis of ecological trauma to become possible. In short, Wallace's pneumatology means that the Earth is "capable of experiencing loss and trauma."[80]

Skeptical Voices

The foregoing survey of possible options for grounding the idea of a traumatized Earth yields mixed results. On the one hand, there are a range of overlapping entities that are all potential subjects of ecological trauma, including the world soul, the *Logos*, a cosmopsychic universe, Gaia, an interconnected and distributed consciousness, a reanimated Earth, and the Holy Spirit. But there are also some major hurdles for these ways of thinking: they may imply too much by way of divine order, they may lack philosophical plausibility, they may run into conflict with Western science, they may offer little by way of connection to Christian theology, or they may threaten divine impassibility. None of these problems are necessarily insurmountable, but they do give a Christian theologian pause for thought.

The deeper issue here is that both Western science and Western theology have tended to be highly skeptical of any form of subjectivity beyond the human. Scientifically speaking, this skepticism has run in parallel with the accomplishments of Western science. If there is no soul or spirit in nature, then science and technology are free to measure, use, and exploit nature's inanimate "resources." The extraordinary achievements of this technoscientific paradigm, including everything from medicine to space exploration, has persuaded Western culture at large that it is self-evidently correct to assume that the planet is devoid of subjectivity.[81] Theologically speaking, this skepticism has relied on a strong distinction between Western monotheisms and pagan animisms. As Lynn White explains in his famous essay on the roots of the ecological crisis: "By destroying pagan animism, Christianity made it possible to exploit nature in a mood of indifference to the feelings of natural objects."[82] A simplistic narrative of Christian replacement leaves no room for nonhuman subjectivities. Indeed, several popular and academic theologians today report anti-animist sentiments as central tenets of their faith: "from Genesis to the present the biblical world-view has clashed with the world-view of animism"; "the whole of the religious tradition of the West . . . [argues] that nature has no soul and is not in the salvific plan of God."[83] Such statements are factually incorrect: the "biblical world-view"—as we have seen—is certainly open to an animist reading, and the "tradition of the West" has regularly returned to the possibility of a world soul. Yet, again and again, the idea of an animate Earth is dismissed as too much like paganism or pantheism. Mary-Jane Rubenstein has done important work to illustrate how this "relentless demonization and name-calling"—especially in the case of pantheism—invariably stems from racialized or gendered prejudices; nature is often cast as dark or feminine.[84] Unfortunately, though, no matter how incorrect or problematic such dismissals are revealed to be, anti-animist assumptions remain deeply rooted in the West. The absence of a soul or spirit in nature seems to have migrated into the realm of "common sense knowledge" to the point that justifications are no longer needed.[85]

In fact, this skepticism is so widespread that it even infiltrates the work of several writers who are actively seeking to recommend various forms of new animism. Roszak, for example, is frank about the extent to which reductive and materialist interpretations of modern science are tacitly accepted within contemporary Western culture. "Science permeates the lives of people everywhere in the modern world," he writes, and so, "hearing 'tongues in trees, sermons in stones' — as anything more than poetic license — has come to be regarded as the very essence of superstition."[86] Likewise, Lovelock is acutely aware that for his theories to be accepted by the scientific establishment they must be "purged of all reference to mystical notions" and must remain "within the strict bounds of science."[87] In her book *Vibrant Matter* Jane Bennett makes an explicit argument for nonhuman agency. However, despite this aim, some of her comments reveal just how ingrained the aversion to animism has become in Western thinking. She describes "discredited modes of thought . . . [such as] animism, the Romantic quest for nature, and vitalism," and wants to avoid being "infected by superstition, animism, vitalism, and other premodern attitudes."[88] Even those who are vigorous supporters of different dimensions of new animist ontologies are nonetheless wary of superstition, infection, and discreditation.

It is also important to highlight how Western modernity's suspicion of animism has developed within a colonial context. The Victorian anthropologist Edward Burnett Tylor coined the word "animism" to characterize what he saw as a primitive and narcissistic "religion of the savages."[89] For Tylor, Indigenous traditions and spiritualities amounted to little more than fetishism and idolatry. He thought that religion in human societies "progresses" from animism, to polytheism, and finally to monotheism. One response would be to seek to overturn Tylor's deeply problematic assumptions by attempting wholesale decolonization of Western thought. For example, anthropologists following the "ontological turn" seek to take emerging metaphysical claims made by Indigenous peoples and cultures much more respectfully and seriously.[90] In one sense, Indigenous thinkers might be amused to see Western philosophy finally catching up.[91] But we should also be wary of thinking that the West can simply appropriate heterogeneous Indigenous perspectives as a panacea for deep-rooted ills. As Walking Buffalo puts it, "trouble is, white people don't listen. They never learned to listen to the Indians, so I don't suppose they'll listen to other voices in nature."[92]

Behavior Before Belief

Mari Joerstad, in her reading of animistic texts in the Hebrew Bible, wrestles with how best to engage Indigenous animisms. Joerstad is candid about how, despite the best of intentions, Westerners can find that reading about animism "does not erase the strangeness of interacting with trees or rocks as subjects capable of response."[93] As some commentators put it, the language of animacy and agency in the natural world is "almost impossible for our modern Western minds."[94] In her opening pages, Joerstad recounts her experience of interacting with two trees, *Fryd* and *Glede*. But when she considers what the trees themselves are experiencing, she concludes that "I don't know and I don't know how to begin to know."[95] There is a striking honesty to Joerstad's agnosticism. For those who have been brought up amid a raft of Western assumptions, it feels like we do not know enough to be able to make any final pronouncements one way or the other about the animacy of the nonhuman. This is not so much an active position that can be argued for as a recognition of epistemological limitation. We simply cannot know what it is like to be a tree, or a planet.

Wallace is a little more optimistic about the prospect of Western societies adopting an animist belief system. He writes, "the urgency of our era necessitates such risks [as Christian animism] in order to explore theologies willing to call ecological violence into question, even at the cost of partnering with frameworks heretofore viewed as heterodox."[96] Wallace's plea is that exceptional circumstances, like the current ecological crisis, legitimize the appeal to some form of new animist spirituality, even in the face of the West's widespread suspicions. Wallace is not wrong about the urgency of the contemporary situation, but the opposite logic may be just as plausible. If the goal is to bring about more ecologically sensitive behaviors, then we need to bring as many people on board as possible, without taking the risk of discouraging them by first requiring them to accept what they perceive to be heterodox modifications of their worldview. Wallace himself effectively concedes as much when he asks "whether such new animism can pragmatically undergird policy decisions."[97] He implicitly acknowledges that insisting on an animist worldview is not a terribly realistic approach within a reductive technoscientific culture.

To be clear, planetary animism need not necessarily be understood as either superstitious or antithetical to Christian theology. Moreover, many animist philosophies contain immense potential for informing ecologically minded modes of living, if only they were to be more widely embraced. Yet the stranglehold of Western suspicion risks undermining the notion of ecological trauma if it depends entirely on an animist ontology for its articulation. In Western scientific and theological contexts, we cannot rely on people drastically modifying their views about the presence of a soul or spirit in nature before introducing a theology of ecological trauma. To speak candidly, my hope is that this book still finds an audience among those Christian theologians who hold reservations about a fully fledged Christian animism. My own havering on the question of Earth's animacy also derives from a sense that ontologies are not as important as they initially seem. Drawing a sharp contrast between animist and modernist *beliefs* distracts attention away from how people *relate* and *behave*.

In fact, a vital thread that links all the positions outlined above is a prioritizing of behavior over belief. One of the primary desires of panpsychists, Gaia theorists, ecopsychologists, Indigenous communities, and animists of all stripes—not to mention biblical authors and Pope Francis—is to relate to the Earth in a caring and sustainable way, and only secondarily to affirm any specific beliefs about the Earth. For Lovelock, Gaia is "the seed from which an instinctive environmentalism can grow."[98] For Roszak, animism has "a proven ecological utility" because it imposes an "ethical restraint upon exploitation and abuse."[99] For Rubenstein, "animacy is not a matter of belief but rather of relation."[100] For Wallace, Christian animism gives us "vision and energy to enter the public fray" and can "empower our collective desire to heal Earth's suffering."[101] And Francis's underlying aspiration is that "we may protect the world and not prey on it."[102] In all of these cases, belief is important to the extent that it undergirds a desired behavior. The action leads to the knowledge, not the other way around.

This combination of agnosticism concerning belief and commitment concerning relationship is helpfully encapsulated by Martin Buber's famous reflections on his interactions with a tree:

> It can, however, also come about, if I have both will and grace, that in considering the tree I become bound up in relation to it. . . . The tree is no impression, no play of my imagination, no value depending on my mood; but it is bodied over against me and has to do with me, as I with it—only in a different way. . . . The tree will have a consciousness, then, similar to our own? Of that I have no experience. . . . I encounter no soul or dryad of the tree, but the tree itself.[103]

Buber remains unsure about the consciousness of the tree, but he does allow for an encounter with the "tree itself," letting the body of the tree impinge upon him and form a relationship. The important element here is the quality and actuality of the relationship that is established, not the belief that Buber holds. Paul Santmire also develops Buber's reflection. Santmire notes how Buber does not advocate for an objectifying I-It relationship with the tree, but neither does he unambiguously support an I-Thou relationship, because he holds that I-Thou relationships are typically predicated on mutuality and spoken communication.[104] Hence, Santmire proposes the introduction of a third category, which he describes as I-Ens relationships. An I-Ens relationship is akin to an I-Thou relationship, but without the literal speech. The Ens (or being) is not subordinated within a utilitarian understanding of the world, but is instead respected in its "givenness," its "mysterious activity," and its beauty.[105] The tree is simply to be beheld and contemplated, rather than used, or protected, or served. Following Buber's example, it is the way in which we are "bound up in relation" that is of most significance.

When it comes to ecological trauma, relationships and behaviors that work on the assumption that the Earth is traumatized arguably matter much more than any specific beliefs about the Earth's subjectivity. And this is a position that is well served by attending to the power of anthropomorphic metaphor.

Ecological Trauma as a Relational Anthropomorphism

Anthropomorphisms have recurred throughout this chapter, from the mourning land in Hosea and the cry of the Earth in *Laudato Si'* to the intelligence of Gaia and the life of the planet. If we remain agnostic about animist ontologies it could be tempting to assume that such formulations are "merely" metaphors in the sense of ornate textual decorations. However, it is an impoverished understanding of metaphor that claims these animate anthropomorphisms are elaborate, but ultimately simply false, tactics for describing earthly processes and phenomena. A much fuller understanding of metaphor recognizes that the semantic innovation involved is part of how we think and how we interact with the world.[106] Metaphor is central to how we relate to the entities around us. As the Earth Bible Team state, "the metaphor of voice is more than metaphor."[107] To put it another way, metaphor has affective and relational dimensions, not just cognitive and comparative ones. This is because our use of language both emerges from and subsequently impacts our embodied experience. As Joerstad explains in relation to the Hebrew Bible, "metaphor allows biblical authors to explore the multiform relationships between humans, nonhumans, and God . . . thereby enabling discovery in the absence of definitional knowledge."[108] With this more robust appreciation of metaphor in hand, it is possible to begin thinking about the traumatized Earth in this vein. We may discover something about the Earth and how to interact with it if we are willing to describe it as traumatized. Yet the anthropomorphism involved means that there are still two serious hurdles to overcome if Christian theologians are to become comfortable with the metaphorical ascription of ecological trauma.

First, since at least Ludwig Feuerbach, theologians have tended to be constitutionally wary of anthropomorphism. In *The Essence of Christianity*,

Feuerbach maintains that theology consists in the projection of human characteristics onto an empty cosmos and the subsequent reification of these projections into ontological entities.[109] The theological risk is that anthropomorphism is really idolatry; it is not God that is being envisaged, but magnified versions of various human needs and desires. Similarly, when anthropomorphic projection is turned on the natural world, the risk is that we are committing the pathetic fallacy: sentimentally attributing human traits to entities that do not possess such capabilities.

One solution, in the case of theological anthropomorphism, is to revert to a strong view of revelation. Christianity is not just wishful projection because the communicative impetus is divine, arising from God's own self-revelation. Likewise, in the case of ecological anthropomorphism, a counterargument to the problem of projection follows a similar model. Humans are not projecting thoughts, feelings, and emotions into empty space. Rather, the natural world is already active; humans are merely projecting onto existing activities. Other creatures and other parts of the natural world are regularly initiating communication, whether this is the more literal screams of specific animals, the chemicals emitted by plants in distress, or the more abstract signs of a warming planet. In each case, the reality of the Earth's communication exists independently of our anthropomorphic description of it.[110] It is the natural world that is taking the initiative in revealing something to us. We are the ones who start out as passive, we "surrender" to our encounter with nature, and we let the natural world be a revelation to us.[111] The communicative event may appear as a "mute appeal," but by rendering this in human speech—by giving the Earth a voice—we are not demanding a linguistic form of communication, but rather opening ourselves up to an encounter beyond our logocentric confines and then attempting to translate that experience into a humanly relatable form.[112] As Hayes writes in relation to scriptural examples of mourning, "the state of the earth itself is held up as a silent witness."[113] The witness is embodied, but inarticulate. We allow ourselves to receive these preexisting realities as if they were human to help us appreciate what is occurring. This is where the anthropomorphism lies.

The second concern with anthropomorphic metaphor is the opposite of the first: not that our projections are ultimately hollow, but that by projecting onto the natural world we are not allowing nature to speak on its own terms. We risk silencing nature by smothering it with human concepts.[114] The criticism here is that anthropomorphism is too anthropocentric; it operates on the assumption that everything in the world works like a human being, or is best understood in human terms. Joerstad also worries that anthropomorphism and personification imply that we first perceive other entities as inanimate objects and only later project human characteristics onto them. As she writes, the Earth "was not inert and then enlivened by means of metaphor."[115] Such an approach does not adequately allow for the animate otherness of the Earth.

These are legitimate concerns, but they need not fatally undermine an anthropomorphic approach to ecological trauma. One important riposte indicates that it is just as myopic to think that human beings are unique and fundamentally incomparable with everything else in the world as it is to think that natural entities are always best understood in human terms.[116] Accusations of anthropocentrism cut both ways. The sort of anthropomorphism that is likely to be beneficial when grappling with ecological trauma is not an arrogant assertion that other creatures and entities are like us, but a humble recognition that, when faced with our own limited imaginative capacities, anthropomorphizing other beings offers a way forward. We can never entirely evade the human condition and so we can never fully inhabit a biocentric or ecocentric perspective. We simply cannot know what it is like to be a tree—as Buber puts it, "of that I have no experience"—but we can find a sympathetic way of relating to the tree by starting from the assumption that it is like a human person.[117] The way that we begin to understand and relate to the radical otherness of nature is by conceiving of its nonhuman modes of communication in human terms. Perhaps, as Bennett suggests, anthropomorphism is even a partial cure for anthropocentrism.[118] Furthermore, the metaphorical connection is not simply unidirectional. As well as anthropomorphizing nature, we regularly project the characteristics of the natural world onto ourselves. We talk, for example, about being as faithful as a dog, as wise as an owl, or as strong as a rock. If we can discover something about ourselves by projection of the other, then we may also discover something about the other by projection of ourselves. Finally, if the concern here is about the possibility of false projections, then the solution is not to reject projection as a mode of understanding, but rather to seek to make "better contact" with the world.[119] If we are to avoid smothering nature in overly human qualities, we need to be more open, not less, to what nature is saying.

Psychological studies of anthropomorphism further bolster this argument. Gabriella Airenti defines anthropomorphism as "the attribution of human mental states or affects to non-humans."[120] Crucially, though, this is not a practice that can be explained away as childish, irrational, or primitive. Such suppositions are the product of the same colonial nineteenth-century cultural anthropology that scorned animism. Rather, anthropomorphism is a healthy psychological practice that is grounded in interaction; beliefs are only appended *a posteriori*. As Airenti writes, "to treat anthropomorphism as a system of beliefs without considering its relational aspect is a source of misunderstanding and potentially contradictory results."[121] Anthropomorphism is also more than mere "metaphorical reasoning" because it affects the world as well as just describing it.[122] "Anthropomorphizing objects or biological entities," says Airenti, "is a means to establish a relation with them, dealing with them as interlocutors in a communicative interaction."[123] Anthropomorphism is therefore an appropriate psychological technique for inhabiting an agnostic middle ground between assertions about animist belief systems and a degeneration into "mere" metaphor.

When it comes to the traumatized Earth, anthropomorphic metaphor is a viable way to conceive of what is going on because it creates an affective relation.[124] For example, we can still hold to the ecopsychological conviction that "our planet is sending us signals of distress," without passing judgment on the planet's state of consciousness.[125] The Earth may not have a literal voice, but its trauma can still be communicated—via the cry of the poor, via the complaints of other species, or via rising sea levels and melting ice caps—and we can still treat the planet as traumatized. Our behaviors should precede our beliefs. In Lauren Woolbright's words, "the human capacity for metaphor accesses nature's silent trauma."[126] If we do not allow for this anthropomorphic relationality, then the risk is that the cry of the Earth will go unheard.

Conclusion

This chapter opened with an account of the mourning land in the Hebrew Bible and the cry of the Earth in *Laudato Si'*. My aim has been to show that these anthropomorphisms are not only deeply rooted in the Christian tradition, but also provide a basis for talking about the traumatized Earth. In reviewing a range of traditions and philosophies that might provide an ontological grounding for ecological trauma—including notions of a world soul, panpsychism, Gaia theory, ecopsychology, and animism—I uncovered several entities that could be the subject of Earth's trauma—such as the world soul, the *Logos*, a cosmopsychic universe, Gaia, an interconnected and distributed consciousness, a reanimated Earth, or the Holy Spirit. Yet, despite these possibilities, Western science and Western theology remain somewhat skeptical about the attribution of subjectivity beyond the human. Some of this wariness undoubtedly stems from a colonial mindset that has sought to downgrade and discredit animist spiritualities. But it nonetheless remains challenging to convince Christian theologians in Western contexts that the Earth should be considered agential and responsive. Crucially, however, what is most important here is not conversion to an animist worldview, but encouraging ways of speaking and behaving that hold the potential for improved relations. As such, ecological trauma is helpfully understood as a relational anthropomorphism: a concession to our limited imaginations that helps to enhance our understanding of, and relationship to, the natural world. A robust understanding of metaphor means that this is not a retreat into mere poetic fancy, but a well-worn path for negotiating real relationships. Although we must be careful not to smother the voice of the Earth, we are not simply projecting human emotions and feelings onto empty space because the communicative impetus is coming from the Earth itself, only in a form that undercuts our expectations of literal speech. This is also an approach that finds support among psychologists, biblical scholars, and theologians, including Pope Francis. If ecological trauma is understood in this way, it is perfectly plausible for a Christian theologian to speak about ecologies, ecosystems, and even the Earth itself as traumatized.

2
Trauma in Ecotheology

> In minding the emerging crisis of the earth, we may name the unassimilable, the irreparable, as ecotrauma.[1]
> —CATHERINE KELLER, *Political Theology of the Earth*

During the early months of 2020, Australia was consumed by fire. There were hundreds of separate infernos, some with fire fronts hundreds of kilometers long. Cities were filled with acrid smoke and several blazes were so intense that they created their own weather systems. In total, some twenty-four million hectares was reduced to ash, including much ancient rainforest that had never burned before. An estimated one billion animals perished in the flames, and several endangered species are believed to have been driven to extinction. Writing in January 2020, in the midst of this unfolding catastrophe, the ethicist Clive Hamilton said that it felt like "the apocalypse has come."[2]

The Black Summer bushfires of 2019 and 2020 constitute one of the most intense and catastrophic fire seasons on record. Typically, less than 2 percent of Australia's temperate broadleaf forests are expected to burn each year. But in the heat and drought of late 2019 and early 2020 more than 21 percent of this biome was incinerated.[3] As is well known, global climate change is making wildfires burn for longer, and with greater intensity.

For some, there are ethical lessons to be learned from the Black Summer fires: behaviors ought to be changed, and practices ought to be adopted that will prevent the recurrence of such disasters.[4] Tangible efforts to mitigate the impacts of future catastrophes are certainly not to be shunned. But, for others, the situation feels existentially acute. Many conservationists saw not only their life's work, but also their optimism about the future go up in smoke. As

Hamilton put it, in the heat of the blaze, these fires are "upending our ways of thinking about the world."[5] There is something ungraspable about both the scale of the damage and the threat that is posed to future ways of life. No ethical introspection or disaster risk reduction is going to alleviate these concerns. As Hamilton continued, "it's hard to know how the trauma will express itself once the fires have burned themselves out."[6] We are dealing here not so much with ethical priorities as with trauma response.

For an ecotheologian, the Australian wildfires present a pressing test case for thinking about current approaches to ecological suffering. In this chapter, I specifically consider how the concept of trauma could augment existing discussions. In other words, how can the vocabulary of trauma—as Hamilton introduces above—help to reframe the way in which an ecotheologian conceives of various forms of ecological suffering? And how might the methodologies of Christian trauma theology provide complementary alternatives to both environmental ethics and natural theodicy? In short, what can ecotheology gain by employing the lens of trauma? In the second half of this chapter, I turn to the few brief mentions of ecological trauma in the extant ecotheological literature to tease out some initial answers to this question. As the climate crisis continues to unfold, the Black Summer bushfires offer just one foretaste of the magnitude and the incomprehensibility of what is to come. My proposal is that the category of trauma helps ecotheologians to grapple with the overwhelming nature of ongoing and recurring catastrophes. In other words, ecotheology would benefit from *a theology of ecological trauma*.

The New Language of Trauma

The way in which we speak about climate change and ecological destruction is changing. Activists have been at the vanguard of this shift in terminology, proposing the language of crisis, chaos, and collapse to underline the severity of the contemporary situation.[7] For instance, Extinction Rebellion talks in terms of "unprecedented" climate breakdown, the "annihilation" of biodiversity, and a "dying planet."[8] But this desire for more urgent vocabulary has also become mainstream. In 2018, United Nations Secretary General António Guterres spoke forthrightly about the "climate crisis." In 2019, *The Guardian* newspaper updated its style guide to recommend the use of "climate emergency" instead of "climate change," and "global heating" instead of "global warming."[9] And in 2021, the *Oxford English Dictionary* added both "climate crisis" and "global heating" to its list of recognized words and phrases.[10] Some—such as the climate scientist and evangelical Christian Katharine Hayhoe—have even invoked the notion of "global weirding."[11] In each case, the purpose seems to be twofold: the rhetorical intensification undoubtedly seeks to galvanize change, but there is also the sense that some of the new vocabulary is more true to the reality being described—a reality that is rapidly spiraling out of control.

For Timothy Morton, the language of "climate change" is both insufficiently demanding and insufficiently precise to capture what we are talking about. The phrase "climate change" does not indicate whether the planet is getting hotter or colder, let alone how fast or with what consequences. It also allows deniers to point to the fact that the climate has always been changing. On these grounds, says Morton, even the outdated-sounding phrase "global warming" is preferable.[12] Moreover, "climate change" has become normalized within social and political discourses. It is one issue among the many that vie for the attention of individuals, corporations, and policymakers. Consequently, there is little recognition that we are talking about something existential: the very ending of life as we know it. A linguistic shift is required, says Morton, because "the phrase climate change has been such a failure."[13] "What we desperately need," they continue, "is an appropriate level of shock and anxiety concerning a specific ecological trauma."[14] But Morton even goes one step further than this. They propose that the very term "climate change" is itself "a kind of denial" and "a reaction to the radical trauma of unprecedented global warming."[15] The language of climate change has allowed us to assimilate the phenomenon of global heating into our discourse without us having to grapple with the seriousness of what is at stake. It is a dissociative defense mechanism that prevents us from engaging with the reality of a radical trauma.

Zhiwa Woodbury recounts a very similar sentiment in relation to our use of language. He writes, "I can no longer, in good conscience, refer to this accelerating threat as 'global warming' or 'climate change.' *Climate Trauma* is emphatically a more descriptive, and notably more useful, term for what we are now experiencing."[16] According to Woodbury, we need to shift from a climate change paradigm to a climate trauma paradigm if we are to bypass the political polarization, institutional failings, and incrementalism that have plagued responses to climate change.[17] Again, the new terminology is thought to be both more effective, and more true to the phenomenon being described.

What Morton and Woodbury reveal is that the vocabulary of trauma is a central component of a wider change in how we describe and relate to ecological emergencies. The word "trauma" seems able to point toward something shocking, radical, and unprecedented. To be blunt, something described as traumatic must be *really* bad. It is the language of trauma that might just shake us from our sociopolitical comfort zones and awaken us to the existential force of ecological destruction.[18] The climate crisis is not just a situation that is slowly worsening with time, but a phenomenon that could catapult the Earth into a new planetary state altogether. Likewise, species extinctions constitute not just unfortunate reductions in overall levels of biodiversity, but complete and irreversible endings of whole ways of life.[19] These are traumatic ruptures, not mere changes. The category of trauma is therefore an important addition to this recent strain of more urgent conversation.

Within ecotheology, a variety of vocabulary has been employed to describe ecological suffering. Scholars mention, for example, the need for theologians to respond to "environmental harms," "environmental concerns," and "environmental issues."[20] In *Laudato Si'*, Pope Francis writes about the problem of "environmental degradation."[21] When referring to ecological suffering that is deemed to be natural rather than anthropogenic in origin, including many instances of species extinction, environmental theologians and philosophers have often spoken about the "disvalues" that are part of the natural world.[22] To be clear, there is nothing factually incorrect about drawing attention to the harms, concerns, issues, and disvalues of a degrading environment, but such terminology can at times feel inadequate for capturing the magnitude and extent of ecological suffering, especially in the midst of the climate crisis. Ecotheologians have therefore had cause to adopt a more insistent tone. Francis, for example, also invokes the language of "ecological crisis" on a regular basis throughout his encyclical letter.[23] And some ecotheologians have begun to speak in terms of ecological trauma (see below).

To reiterate, trauma terminology has entered ecological discourse as part of a wider movement to intensify the rhetoric around climate change and environmental destruction. But this shift to the new language of trauma is more than just a communication strategy. There are also various ways in which the concept of trauma is appropriate for diagnosing the phenomena at hand, serving as a placeholder for forms of ecological suffering that are overwhelming and irreversible. Trauma deserves to be engaged ecotheologically.

Environmental Ethics and Natural Theodicy

Existing ecotheological approaches to ecological suffering tend to divide into environmental ethics on the one hand, and natural theodicy on the other.

Christian environmental ethics usually operates with the implicit (or sometimes explicit) assumption that the ecological suffering in question is the fault of human beings. As Michael Northcott puts it, "the environmental crisis is rooted in moral evil and human guilt."[24] It is not hard to see how such a conclusion is reached. The ecological devastation entailed by the environmental crisis—including suffering, death, and extinction—clearly results from human hunting, felling, clear-cutting, quarrying, extracting, mining, polluting, emitting, and so on. Indeed, several Christian ecotheologians have appealed to the concept of "ecological sin" as a framework for registering humanity's damaging attitudes and activities.[25] In Northcott's words, the problem is "the turning away of modern human civilisation from God's order and redemptive purposes for the world."[26] In addition, Christian ethicists must wrestle with the suggestion that Christianity is not just a moral guide on these issues, but also part of the problem. If the Christian worldview promotes an anthropocentric and dominating mindset, this can legitimize an extractive and exploitative use of the natural world.[27] Either way, whether Christianity is to blame or not, the environmental ethicist understands ecological suffering to result from human wrongdoing. The solution to ecological suffering is a closer examination of ethical precepts and norms, coupled with a renewed resolve to implement the resulting moral guidelines. For Northcott, this involves returning to an environmental ethics based on natural law.[28] Alternatively, in the case of the Australian bushfires, this could be about learning from Indigenous communities and moving toward "a new ethic of nurture, sustainability, and respect."[29] In general, Christian environmental ethics "has the task of formulating orientations for good and just action and critically reflecting on the various proposals for this."[30] In each case it is humanity that has the power, and therefore the responsibility, to respond.

By contrast, natural theodicy addresses examples of ecological suffering that are not thought to be the fault of human beings. Traditionally, natural theodicists deal with ecological suffering that is said to stem from "natural evils" such as predation, disease, and natural disasters. Human beings are not usually held responsible for such phenomena, not least because these forms of ecological suffering all predate the biological evolution of the human species. Other creatures were afflicted by predation, disease, and disaster long before the arrival of humanity. We are not to blame—whether via a theology of Adam's fall, or anything else—for suffering that occurred before our time.[31] Instead, natural theodicists seek to explain why ecological suffering is necessary for some greater good, or how a good God can allow it to occur. They tend to argue that natural evil is the result of a mysterious angelic fall; that it is indispensable to the fruitfulness of the evolutionary process as a whole; that it is an unavoidable by-product of a universe fine-tuned for life; that all suffering creatures will receive eschatological compensation; or that it is in God's essence to be loving, self-giving, and hence self-limiting.[32] For example, in relation to the Australian bushfires, one might make the argument that a certain amount of burning is both natural and necessary for clearing thick undergrowth, promoting seed germination, eliminating disease, and benefiting biodiversity. In these cases, reasons are given—whether scientific of theological—to justify the ecological suffering that occurs.

Both environmental ethics and natural theodicy have their place, but there are also some potential problems with these approaches to ecological suffering. This is where the category of trauma, and a trauma-informed methodology, can offer ecotheologians a third way.

Beyond the Nature/Culture Dichotomy

Within ecological circles it is widely recognized that a sharp division between human culture on the one hand, and the rest of the natural world on the other, rings increasingly hollow.[33] It is simply not the case that human activities take place in isolation, with nature merely providing a passive backdrop. Instead, human and natural processes are constantly interacting and interfering. As Bruno Latour writes, one only has to read the news to be reminded that "all of culture and all of nature get churned up every day."[34] For Latour, this is not to deny that there are some differences between the human and the nonhuman, but rather a reminder that divisions of "the seamless fabric of . . . 'nature-culture'" are always somewhat arbitrary.[35]

Yet, in terms of ecotheological approaches, any sharp division between environmental ethics and natural theodicy perpetuates the nature/culture dichotomy. According to this distinction, ecological suffering is understood to be the result of either moral evil or natural evil in a zero-sum game. Deforestation, opencast mining, and anthropogenic climate change are considered moral evils to be addressed by environmental ethics because they are assumed to originate "outside" nature; whereas viral pandemics, volcanic eruptions, and devastating earthquakes are considered natural evils to be addressed by theodicy because they are assumed to originate "within" nature. It is a division that stems from a worldview in which human actors are understood to be entirely detached from their surrounding environments. But such separations ignore the churning and interacting agencies that Latour describes. For example, nature plays its part in anthropogenic climate change via environmental feedback loops that worsen the warming effect of the initial emissions. Likewise, humans play a part in determining the destruction caused by an earthquake, both because activities like fracking can sometimes trigger small seismic events, and because the location and integrity of buildings can have a significant impact on the death toll. To return to the Australian bushfires once again, the fires have a natural cause inasmuch as lightning strikes are often the trigger for a blaze, and the burning process is part of a natural cycle of regeneration. But the fires are also human in origin since failures to manage the bush allowed the fires to spread rapidly, and anthropogenic climate change created the hot and dry conditions that significantly intensified the inferno. Moral evil and natural evil are not quite so easily disentangled.[36]

One way around this brittle binary is to propose some additional categories. Celia Deane-Drummond, for example, suggests that we augment moral evil and natural evil by considering the possibility of "anthropogenic evil"—that is, evil and suffering in the nonhuman world caused by human actions such as pollution and habitat destruction.[37] This definition recognizes the impact that human activities have on natural processes, but it does not quite capture the ecological suffering that results from the interaction of human and nonhuman actors. The natural and the cultural are still presupposed as separate entities.

More generally, environmental ethics and natural theodicy both direct attention toward the *causes of* and *solutions to* ecological suffering. In both disciplines the operative assumption is that the primary task of the ecotheologian, when faced with ecological suffering, is to categorize it based on its origin and then supply some sort of explanation. In many cases, diagnosing the cause of ecological suffering is helpful precisely because it indicates the sort of solution that is likely to be effective. If the Black Summer bushfires were caused by poor management practices and inadequate fire breaks, then processes can be improved to reduce the likelihood of future catastrophes. But, given the entanglements that complicate the nature/culture dichotomy, it is often challenging to ascertain who exactly is to blame for a given instance of ecological suffering. Likewise, the character of the suffering may well mean that no easy solution is available.

This is where Christian trauma theology offers the ecotheologian an alternative approach. The methodologies embedded in trauma theology promote strategies for learning to live with, rather than categorize or solve, traumatic forms of ecological suffering. Trauma theologians recognize that instances of trauma can have human causes, nonhuman causes, or both. But they do not depend on being able to pinpoint the type of evil involved before they can offer a response. As Serene Jones is acutely aware, it is a common and entirely understandable human impulse to be looking for causes and solutions. "My almost instinctual response," she says, "is to start trying to figure out what went wrong and then to start looking around for what will fix it."[38] But Jones goes on to note the frequent failure of this impulse. As she writes, "the vast majority of trauma survivors reach the end of their lives still caught in its terrifying grip."[39] In many instances, a persistent search for causes and solutions does little to help with the actuality of living with ecological suffering in the present. Trauma theology therefore encourages the ecotheologian to bracket causes and solutions in favor of an approach that seeks to console and accompany. As Shelly Rambo explains, trauma theology is best done from the "middle," operating in the midst of ongoing suffering, without knowing what caused such pain or how it might be resolved.[40] Instead of seeking to silo instances of ecological suffering based on whether such suffering originates "outside" or "within" nature, an ecological trauma theologian offers attention and care regardless.

Beyond Rational Control

Environmental ethicists typically agree that "rationality is an essential ingredient" of their approach to ecological suffering.[41] As Markus Vogt comments, Christian ethicists are "committed to the claim of rational argumentation without any restriction."[42] But this commitment to rationality can also present environmental ethicists with a problem when they confront an issue on the scale of the current ecological crisis.

As Hamilton argues, conventional ethical thinking often feels inadequate to the task when it comes to responding to climate change. The unparalleled scale and scope of contemporary ecological devastation results in what might be termed the "banality of ethics in the Anthropocene."[43] What Hamilton means is that "when we step back and survey these Earth-shattering events our established ethical categories . . . appear banal and feeble."[44] To label a specific activity such as commercial deforestation or raw sewage discharge as merely "unethical" is a category mistake. Such events are no longer just grist for the ethical mill; they are contributing to setting the mill on fire. Our response to the ecological crisis is not merely one more ethical decision among others; it is about the future of the planet and therefore the very context in which all other decisions are made. As Hamilton writes:

> It is not enough to describe as "unethical" human actions that are causing the sixth mass extinction of species in the 3.7 billion-year history of life on the planet. The attempt to frame such an overwhelming event by mere ethics serves to normalise it, of reducing it to just another "environmental problem." Talk of ethics renders banal a transition that belongs to deep time, one that is literally Earth-shattering.[45]

In other words, there is a risk that applying rational ethical frameworks to contemporary ecological breakdown can lead to the misapprehension that human beings have the situation under control when the reality is, as Hamilton says,

"Earth-shattering." A model in which a natural scientist describes the issue, and an ethicist says what ought to be done about it, will not suffice because the problem cannot be captured in such a straightforward way. Shared ethical norms are simply too weak in relation to the permanent losses of species extinctions and the irreversible impacts of climate change.

By contrast, trauma theorists and trauma theologians might just be capable of pointing toward the Earth-shattering events that Hamilton thinks ethicists are struggling to apprehend. The concept of trauma can potentially perform this function precisely because it has always sought to gesture toward the inarticulable, to that which transcends mere suffering. As trauma theorist Cathy Caruth puts it, "trauma is the confrontation with an event that, in its unexpectedness or horror, cannot be placed within the schemes of prior knowledge."[46] This seems to be exactly what Hamilton has in mind when he speaks of the banality of environmental ethics. In the case of the ecological crisis, a diagnosis of trauma is appropriate because there is something at the heart of the crisis that we seem unable to understand, let alone control. We struggle to grasp the scale of the suffering involved and the implications for the future of life on Earth. Although not currently predicted, it is possible that we might even be instigating our own extinction. Importantly, the diagnosis of ecological trauma is also an indication of the type of treatment that we may reasonably expect. We are not dealing with phenomena that are necessarily solvable or fixable according to any rational calculus. Ecological trauma is something that we must cope with, and learn to live alongside, rather than something we can control or eradicate. We must develop theological strategies for incorporating the incommensurate.

Natural theodicists also tend to deal in rational explanations. The aim is usually to justify, or rationalize, the occurrence of ecological suffering. But many writers, including some of those who offer theodicies themselves, sound a note of caution. For example, Christopher Southgate highlights the "shortcomings of all such theological enterprises," which, he says, must always "arise out of protest and end in mystery."[47] Similarly, Deane-Drummond comments that theodicy is "far too theoretical to really deal properly with the devastation that suffering brings."[48] Indeed, the burgeoning literature on anti-theodicy refers to its lack of pastoral sensitivity, its inability to speak from the perspective of the survivor, and the risk that it conceals rather than addresses the reality of evil.[49] In an ecological context, for example, a reassurance that a given creature died for the good of the ecosystem as a whole offers no solace to the creature in question. Rational justifications have limited utility.

Wendy Farley therefore recommends moving away from theodicy and toward what she calls a "tragic vision." Farley believes that the theodical project of explaining suffering can be deeply unhelpful if it fails to reckon with those radical forms of suffering that are "irredeemably unjust."[50] Justifying the unjustifiable only makes the situation worse. For Farley, "desire for justice" and "anger and pity at suffering" should replace "theodicy's cool justifications of evil."[51] Whenever theodicy's rational explanations begin to crumble, we should turn to an approach that foregrounds passion and compassion. Traditional natural evils—such as predation, disease, and disasters—may legitimately be described as tragedies. But a further advantage of Farley's turn from theodicy to tragedy is that it also incorporates those phenomena that are usually categorized as moral or anthropogenic evils. Again, the nature/culture dichotomy breaks down. Farley does not explicitly address ecological suffering, but John Foster makes a persuasive case that the current ecological crisis constitutes an "environmental tragedy."[52] It is tragic in the proper sense of the word because the devastation is fatally intertwined with the otherwise positive phenomena of technological creativity and the desire for progress. Past devastation is irreversible and future devastation is unavoidable.[53] "Tragedy of its nature entails radical breakdown, with terrible loss that cannot be mitigated or compensated," writes Foster.[54] But there is also relief to be had in admitting these tragic losses; the aim is to generate compassion for those who suffer.[55] On this understanding, we should be focusing as much, if not more, attention on accompanying survivors as we do on offering rational explanations.

Trauma theology offers the ecotheologian an alternative to natural theodicy by adopting exactly this tragic sensibility. This is because tragedy and trauma both "make space for violence and suffering without providing an easy explanation as to its cause."[56] Trauma theologians are used to working with tragedy, lamenting what is irretrievably lost, offering compassion to survivors, and delivering a protest that seeks to reduce future violence. Trauma theology is also attuned to the concerns of the anti-theodicists. As Rambo notes, theodicies "do not function effectively to address and respond to suffering" because it is unclear whether their proffered explanations are any practical help.[57] Trauma theologians seek to *witness to*, rather than *explain away*. And this is a strategy that may help with the irrationality and incomprehensibility of much ecological suffering as well.

To be clear, neither environmental ethics nor natural theodicy are obsolete. Each has its place, and each continues a laudable desire to understand. Moreover, plenty of thinkers within these subdisciplines already recognize some of the limitations described above. For example, Donna Orange is well aware that none of our existing environmental ethics have come close to inspiring the large-scale changes that are needed, which is why she recommends a "radical ethics" that confronts us with responsibility rather than recommending a strategy.[58] Likewise, Bethany Sollereder is attuned to the need for theodicies to do better at engaging directly with the sufferer and become more aware of pastoral and practical concerns, which is why she proposes a "compassionate theodicy" that allows the survivor to construct their own explanations.[59] But an ecological trauma theology offers a parallel, and different, way forward. Environmental ethics foregrounds the role of human beings, and natural theodicy seeks to justify God, whereas trauma theology places the suffering itself at the heart of the analysis. The primary concern is about responding to and living with suffering. This is not to say that human beings, or God, have nothing to do with ecological suffering—for a Christian ecotheologian, they manifestly do—but it is to let the subjects of ecological suffering drive the theological agenda.

Trauma in Existing Ecotheologies

Having proposed trauma theology as an alternative approach to ecological suffering, I now pivot to some more specific diagnoses of ecological trauma within the ecotheological literature. There have, to date, been no extended discussions of ecological trauma by Christian ecotheologians. But there are a few nascent suggestions from Pamela McCarroll, Catherine Keller, Danielle Tumminio Hansen, Mark Wallace, Matthew Eaton, and a few others. Many of these mentions are comparatively brief, but it is worth unpacking the hints they provide for developing theologies of ecological trauma.

Relinquishing Control

McCarroll's investigation of the possible links between trauma theory and the climate crisis is representative of a growing number of pastoral and practical theologians who are beginning to speak in terms of ecological trauma.[60] She proposes that the intersection between trauma and climate is pertinent to pastoral theology because it implicitly relates to existential, theological, and spiritual questions.[61] McCarroll's primary focus is on ecological trauma experienced by human beings, but she does also leave space for forms of collective trauma and Earth trauma too.

For McCarroll, the climate crisis is precipitating a spiritual crisis for humanity because it involves recognizing our own lack of control. As we come to appreciate the sheer extent of ongoing ecological destruction we are forced to confront our own fragility, finitude, and vulnerability—what McCarroll calls our "creatureliness."[62] Viewing ecological suffering through the lens of trauma involves human beings learning how to relinquish their desire for control. Drawing on Elizabeth Stanley, McCarroll suggests that part of the human desire to control ecological suffering is itself a form of trauma response. Confronted with creaturely vulnerability, and overwhelmed by the magnitude of ecological suffering, human beings naturally attempt to categorize and colonize to regain a sense of mastery. McCarroll notes how certain "theologies of glory"—theologies that mask creatureliness with talk of power and victory—are symptomatic of this impulse.[63] But they ultimately only succeed in concealing our deeper fears and occluding ecological trauma. According to McCarroll, "motifs of human and divine power/control" are very often traumatic responses that are trying to "overcome the realities of existing."[64]

McCarroll's insistence on relinquishing control also poses a challenge to existing ecotheological frameworks. According to many ecotheologians, human beings are tasked by God with stewarding and caring for the rest of creation.[65] But McCarroll views this as another trauma-reactive response in which human beings exaggerate their own authority as an antidote to their fear of vulnerability. An ecotheology of stewardship "inverts the deeper truth," says McCarroll, "that humans are actually dependent on and recipients of the earth's stewarding care."[66] The logic of McCarroll's argument also chimes with some of the concerns raised above about both environmental ethics and natural theodicy. Inasmuch as these approaches rely on neat categorizations and rational explanations, they are likely to founder when confronted with ecological suffering that is irrational or uncontrollable. By contrast, theologies of ecological trauma proceed by relinquishing the desire for control.

Unassimilable and Irreparable

Keller mentions the possibility of ecological trauma in passing.[67] "In minding the emerging crisis of the earth," she says, "we may name the unassimilable, the irreparable, as ecotrauma."[68] What Keller recognizes is that there are components of ecological suffering that both exceed our comprehension, and escape any of our attempts at repair.

There are two aspects to Keller's diagnosis here. First, as mentioned above, trauma is a placeholder for precisely that which transcends our categories of understanding. It is a pointer to the inarticulable. There is both an epistemological limit to what can be known and a linguistic limit to what can be said about cases of ecological trauma. This is a consciously and necessarily apophatic approach. Yet, what the language of trauma does enable is a gesture of recognition for what would otherwise go unacknowledged as "unspeakable loss."[69]

Second, Keller's invocation of ecotrauma highlights a condition with no obvious remedy. As I argued above in relation to environmental ethics, this is not something that can simply be fixed. Instead, Keller points toward the way in which she thinks theologians ought to respond by drawing on Rambo's trauma theology. She describes Rambo's approach as follows:

> In the light of trauma theory, she [Rambo] observes the lack of any final fix, any assured future, but a way in the aftermath of loss to "improvise along the tangled lines of what it means to remain." She refuses a triumphalist leap from crucifixion to resurrection; and so she rescripts the very scene of the resurrected messiah.[70]

The pronouncement of ecological trauma is a recognition that theology cannot provide any quick fixes for ecological suffering and should be wary of any simplistic progression from trauma to healing—from crucifixion to resurrection. "Trauma theory in the late twentieth century," says Keller, "demonstrated that suffering does not follow a time line of repair."[71] It is not the case "that *in time* one can just get over or beyond it."[72] Instead, the theologian's task is one of improvising, remaining, and rescripting.

Keller's introduction of ecotrauma is fleeting, but she reiterates two important concerns for theologies of ecological trauma. Theologians must avoid the temptation to say too much, since the recognition of ecological trauma is an acknowledgment of the unassimilable; and theological trajectories of salvation and healing need to be substantially reconfigured, given that ecological trauma names that which is irreparable.

The Traumatized Body of God

Tumminio Hansen proposes that Sallie McFague's metaphor of the world as the body of God enables a conception of the ecological crisis as a trauma experienced by the Earth.[73] "In what ways," asks McFague, "would we think of the relationship between God and the world were we to experiment with the metaphor of the universe as God's 'body'?"[74] Tumminio Hansen offers one conceivable answer by focusing on various parallels between sexual violations of human bodies and equivalent abuses of the Earth. Just as human bodies can be dominated, subjugated, and controlled, the same could be said of the Earth. Given that sexual abuse is very often traumatic, "then perhaps there is a traumatic dimension to the earth's violation" too, says Tumminio Hansen.[75] McFague's body metaphor therefore helps to advance the idiom of trauma for describing the realities of ecological devastation.

Tumminio Hansen's use of McFague's metaphor also brings to the fore the question of God's involvement with instances of ecological trauma. Within this model, planetary traumas inflicted on the Earth are also traumas that are inflicted on God. The advantage of positing such a connection is that it reinforces the seriousness of ecological trauma: violence done to the Earth is also violence done to God. It also underlines the extent of divine solidarity with ecological suffering. Creaturely suffering involves the suffering of part of God's own body, so there is no way that God can be imagined as distant or uninvolved.

But assertion of this equivalence also generates several further questions that need addressing. If the Earth is subject to traumatic violence, does this mean that God is traumatized too? And if life on Earth is destroyed, will this result in the death of God? As Tumminio Hansen asks, "if we are raping the earth, then are we raping God? What would be at stake were God—and not just the earth—vulnerable to this kind of trauma?"[76] Such questions not only sound provocative but also raise doctrinal issues about God's capacity to suffer and the character of the God-world relationship. However, God experiencing trauma is not beyond the realms of theological possibility; plenty of theologians support divine passibility.[77] And McFague herself is cognizant of the implications. As she writes, "the world as God's body may be poorly cared for, ravaged, and as we are becoming well aware, essentially destroyed, in spite of God's own loving attention to it."[78] Furthermore, McFague's panentheist metaphysics—in which the whole universe, and not just the Earth, is viewed as being within God—ensures that God is not fatally compromised by the trauma of the Earth.

For an ecological trauma theologian, the key question here is how exactly to imagine divine presence and divine response in relation to ecological trauma. Several answers are feasible, and the second half of this book considers this issue in detail. Envisaging the Earth as the traumatized body of God is just one possibility.

Wounded Spirit

Wallace broaches the prospect of ecological trauma in his pneumatological take on the ecological crisis. Wallace's ecological pneumatology centers on the conviction that God is present to the whole of creation in the embodied form of the Spirit. As he writes, "the earth is the body of the Spirit . . . [and] the Spirit ensouls the earth."[79] Wallace justifies this statement by pointing to plentiful scriptural examples in which the Spirit is embodied as water, wind, fire, and birds. He argues, for example, that at Christ's baptism the Spirit took on avian flesh, arriving in the form of a dove.[80]

Much like Tumminio Hansen, Wallace's theology is dealing with the nature of divine involvement in ecological trauma. If the Spirit indwells the Earth, as Wallace suggests, then this places the Spirit at risk of serious harm in the context of the ecological crisis. "If Spirit and Earth mutually indwell one another," writes Wallace, "then it appears that God as Spirit is vulnerable to serious loss and trauma insofar as the earth is abused and despoiled."[81] He continues, "the Spirit's suffering from persistent environmental trauma engenders chronic agony in the Godhead."[82] For Wallace, not only is the possibility of ecological trauma urgent and persistent, but it also strikes at the heart of his doctrine of God. Ecological trauma is borne pneumatologically into the Godhead. Like Tumminio Hansen's reading of McFague, Wallace envisages God—this time as Spirit—to be the subject of ecological trauma.[83] As before, this is a potentially appealing avenue for a theology of ecological trauma because it foregrounds God's wholesale commitment to compassion for the Earth, but it also opens up serious questions about the relationship between creator and creation and the possibility of divine passibility.

What is especially interesting, though, is that Tumminio Hansen and Wallace both highlight the merits of a specifically Christological engagement with ecological trauma. Divine involvement with traumatic ecological suffering might best be thought about, not in terms of God the Father or God the Spirit, but God the Son—since it is widely accepted that the divine both became flesh and suffered traumatically in Christ. Christological language already provides a vocabulary for thinking through how the irreparable harm of suffering and death can be conceived as a part of God. For instance, Tumminio Hansen notes that Christology has long wrestled with the question of how Christ experiences "embodied vulnerability" without "compromising the divine essence."[84] Traditionally, this is achieved by invoking Chalcedonian Christology: Christ takes on suffering in his human nature, but not his divine nature. Tumminio Hansen's proposal is that it may be possible to "make a similar argument for the first person of the Trinity."[85] God the Creator could suffer in an embodied form without putting the divine nature at risk. In parallel, Wallace makes the same suggestion about the third person of the Trinity.[86] In his ecological pneumatology, he suggests that the Spirit relates to the world in an analogous manner to the coinherence of the divine and the human in Christ. The relationship between the wounded Spirit and the world can be parsed in Chalcedonian terms.

The important point here is not so much the details of two-nature Christology, but rather the intuition that a Christological framework is helpful for a theology of ecological trauma. Wallace makes this even more explicit when he writes about the traumatic abuse of the Earth being mirrored by the traumatic crucifixion of Christ:

> *Our forebears executed God's innocent son at Calvary in a paroxysm of rage and violence; we do the same by crucifying God's winged Spirit on the Earth through market forces and habitat destruction.* God is crucified afresh when we lay waste to the carnal presence of God on Earth. The paschal trauma of the cross is daily reactualized through our regular assaults on the good creation God has made. The Earth has become *cruciform*: the scars of Golgotha are everywhere. Jesus' crucifixion wounds are now reopened as the whole Earth bears the marks of eco-catastrophe.[87]

It is a visceral and memorable scene. Wallace highlights here an important parallel between the "paschal trauma of the cross" and the trauma being wrought on the Earth.[88] Indeed, this Christological framing of ecological trauma is central to my argument in the second half of this book. Furthermore, much like Keller, Wallace endorses Rambo's "middle discourse" theology as an appropriate theological response.[89]

Eschatological Lament

Lastly in this series of vignettes, Eaton's reference to ecological trauma develops the temporal dimensions of the discussion. When Eaton employs the language of trauma, he wants to emphasize that any prospect of healing or hope for the future sits alongside the reality of continuing ecological devastation. Even in the eschaton, ecological trauma is neither erased nor forgotten: "the pain of ecological trauma," he says, "will be lamented eschatologically."[90] It is not that the violence is endless, but that the consequences of violence can be permanent. For Eaton, this insistence does not stem from a desire to be pessimistic, but from an urge to understand healing as something other than erasure, because to do so would be to betray the violence by forgetting it. "We cannot hope to forget the unforgettable, lamentable violence done to this world," he says, since "failing to acknowledge violence betrays the dignity owed those—human or otherwise—who suffer."[91] Justice, including ecological justice, will not be achieved unless violence is remembered and lamented. Eaton's focus on eschatological lament therefore extends Keller's understanding of ecological trauma: it is not only that ecological trauma is irreparable in the present moment, but also that it persists throughout time and reverberates eschatologically. Ecological trauma points to a permanence that theologians are forced to acknowledge.

Eaton writes in response to *Laudato Si'*, and he notes how Francis is open to the possibility that we may fail to curb anthropogenic violence toward the natural world.[92] He is critical of the encyclical when Francis seems to appeal to a *Deus ex machina* form of eschatological redemption.[93] But Francis does specifically acknowledge the irreparability of ecological devastation in the case of species extinction, lamenting "species which we will never know, which our children will never see, because they have been lost for ever."[94] In other words, Francis recognizes that the consequences of ecological trauma can be permanent.

Moreover, Eaton envisages this eschatological ecological trauma as being taken up into God. Just as for Tumminio Hansen and Wallace, ecocide overlaps with deicide since "the cruciform being of God directly suffers the same pains and trauma experienced in the more-than human world."[95] In some sense, divinity itself experiences ecological trauma. Eaton's mention of cruciformity also reinforces the intuition that Christ's crucifixion is likely to be important for theologies of ecological trauma.

Conclusion

In this chapter, I have sought to show how the category of trauma has begun to seep into ecotheological conversations, and to explain why this new vocabulary is beneficial. I began with the Black Summer bushfires of 2019 and 2020, partly because this fire season has legitimately been described as traumatic, but also because these fires begin to reveal some of the weaknesses of existing approaches to ecological suffering. Both environmental ethicists and natural theodicists have been stunned by the scale and the scope of the suffering that is being wrought by ecological breakdown. Neither approach is invalid, but trauma theology offers the ecotheologian a fresh perspective on these concerns by focusing less on categorizing and solving harms and more on accompanying and consoling survivors. The new language of trauma both recognizes the urgency of ecological catastrophe and serves as a reminder of the irreversibility, the irresolvability, and the incomprehensibility of much ecological suffering.

In the second half of the chapter, I started to sketch the contours of what a theology of ecological trauma might entail by engaging five short cameos from the existing literature. To date, ecotheological treatments of ecological trauma have been comparatively brief. But these examples nonetheless indicate some of the directions that could be taken. McCarroll emphasizes how the lens of trauma involves relinquishing theologies of control. A diagnosis of ecological trauma is also a recognition of the limitations of human power and human reason. In a similar vein, Keller highlights the irreparability of ecological trauma. Ecological trauma theologians must be prepared to acknowledge suffering that is not readily fixed, and reexamine trajectories of salvation in the process. Tumminio Hansen and Wallace investigate the divine relationship to ecological trauma by envisaging the Earth as either the body of God or indwelt by God's Spirit. But both writers also indicate that a Christological approach to ecological trauma is likely to be theologically fruitful. Lastly, Eaton reinforces the permanence of traumatic ecological suffering by insisting on its persistence in the eschaton. Undoubtedly, multiple theologies of ecological trauma are possible. But what these authors all indicate is the viability and the potential of thinking in this direction.

3
Ecology in Trauma Theology

> The different forms of violence I was describing have also been perpetrated against the earth itself. We are witnessing the violation of the integrity of creation. It's as if we're living in a traumatized physical environment.[1]
>
> —SERENE JONES, *Trauma and Grace*

Serene Jones senses the need for a theology of ecological trauma. She is acutely aware that traumatic violence has been "perpetrated against the earth itself." Instinctually, she recognizes that creation has been subject to traumatic violations. It is as if, she says, "we're living in a traumatized physical environment." What Jones intuits is that there is something about the Earth—something precipitated by contemporary climate chaos and wider ecological breakdown—that deserves the attention of trauma theologians. This is a clarion call from one of the leading writers on trauma theology about the need to engage with the Earth.

The epigraph above is from the introduction to the second edition of Jones's essay collection *Trauma and Grace*. But the quotation from Jones continues:

> It's as if we're living in a traumatized physical environment. I know how to name that as a trauma, but I don't know how to talk about this trauma. Most of the literature [on] trauma focuses on its psychological effects and bodily effects, but I don't know how to talk about it in nature, which we don't personify. Nature is a complex system. There is so much more work that can be done in theology and religious communities about where we're going as a human species, which can be aided by understanding trauma.[2]

Jones is convinced that there is much work to be done in theology, including work about the future of humanity, that is going to require a far greater appreciation of the relevance of trauma, and especially the possibility of ecological trauma. But there is a catch: Jones says that she does not know how to talk about it. Her stumbling block is that, according to the literature, trauma only applies to bodies with psychologies, to human beings that have a reliable seat of subjective experience, and she does not believe that nature possesses such a psychology. Yet Jones hints at an answer to her own conundrum when she talks about the personification of nature. Contrary to what she suggests, we often describe nature in human terms. As I showed in Chapter 1, theology is full of such personifications, from biblical examples of the mourning land to Pope Francis's invocation of the cry of the Earth. We can speak of the traumatized Earth as part of a relational anthropomorphism: that is to say, we can treat the Earth as traumatized to improve how we relate to it. There is therefore scope for moving beyond Jones's impasse.

In this chapter, I press forward with Jones's initial intuition about the need for trauma theologians to pay attention to the Earth. I argue that, to do justice to the rapidly evolving and increasingly interconnected understanding of trauma in contemporary scholarship, trauma theology needs to look beyond its anthropocentric horizons and engage with the ecological realm. In part, this is an argument about what trauma theologians have not previously done, what they could do, and what they might do in the future.[3] At the same time, ecological concerns are not foreign to the work of trauma theologians, and I proceed to identify various ways in which ecological dimensions are already latently present in existing trauma theologies. My contention is that these dimensions deserve to be teased out and consciously developed. Not every trauma theologian needs to be an ecotheologian. But the existence of these ecological dimensions within the motifs already employed by trauma theologians provides further evidence of the potential fruitfulness of *a theology of ecological trauma*. In essence, my proposal mirrors that of the previous chapter: while Chapter 2 made the case for the addition of trauma to existing ecotheologies, the aim here is to show why an ecological dimension is needed in current trauma theologies.

To be clear, any underdevelopment of the notion of ecological trauma is no kind of failure on the part of previous scholars. The early trauma theologians were already breaking significant interdisciplinary ground by staging a conversation between trauma studies and theology. The prospect of ecological trauma is simply a fertile avenue of thinking that theologians have not yet fully explored.

The Expansion of Trauma Theology

In recent decades, trauma studies has significantly expanded in popularity, output, and scope. Scholars have come to recognize that trauma is just as possible in instances of persistent and endemic distress as it is following extreme yet isolated catastrophes. For example, in the 1990s, the American Psychiatric Association officially recognized life-threatening illnesses as possible causes of traumatic stress.[4] Trauma has also come to be understood as something that can arise internally as well as externally. Not all traumatic suffering stems from the imposition of external violence; some threats emerge within the traumatized subject.[5]

Arguably the most significant shift, though, is the fact that trauma is being attributed to a growing range of entities. Two examples illustrate this trend. The possibility of "collective trauma" was first proposed by the sociologist Kai Erikson. Erikson spent his career examining the social ramifications of catastrophic events, and he points to a "striking," "resonant," and "haunting" series of symptoms that are shared across many disaster scenarios.[6] He writes, "by *collective trauma* . . . I mean a blow to the basic tissues of social life that damages the bonds attaching people together and impairs the prevailing sense of communality."[7] Erikson's collective trauma includes structural wounds in a community, such as the loss of medical professionals following a disaster in a hospital, and a less tangible mood or ethos, that is then manifest as public statements, cultural productions, or rituals of remembrance.[8]

Meanwhile, several scholars, and especially animal rights activists, have been pushing for the recognition of trauma in other species. For instance, Gay Bradshaw draws together multiple lines of evidence for the existence of psychological trauma in elephants. Bradshaw shows, point by point, how a set of young South African bull elephants who were forced to observe the culling of their family members as juveniles, and were then relocated to unfamiliar environments, fulfill the criteria for a diagnosis of trauma.[9] Bradshaw also circulated the case history of "Jenny"—a middle-aged African elephant who had witnessed the culling of her family, and was subsequently moved to the United States to work first in the circus and then at Dallas Zoo—to five mental health professionals, without any indication of her species, and all five diagnosed Jenny with some form of PTSD.[10] Similarly, work by Stacy Lopresti-Goodman and others has shown that chimpanzees are liable to PTSD when subjected to "maternal deprivation, social isolation, intensive confinement, and repetitive invasive procedures" during laboratory research.[11] Furthermore, rodents have long been used as experimental models for human PTSD, implying that they too are capable of experiencing trauma.[12]

The phenomena of collective trauma and multispecies trauma are significant because they pave the way for a recognition of ecological trauma. Both phenomena decenter the human individual and open up the possibility that trauma might apply to a much wider variety of entities. As Stef Craps puts it, it is time to "challenge trauma studies to move beyond human exceptionalism," and these wider understandings of who or what can experience trauma do just that.[13] It is also noteworthy that this expansion in the field of trauma studies results from a degree of semantic relaxation in the definition of trauma, as well as from new scientific discoveries. Collective trauma is an example of the former: it has become possible to diagnose whole societies as traumatized because trauma is now understood to refer to more than just physiological and psychological processes. Whereas multispecies trauma is an example of the latter: elephants, chimpanzees, and rodents are believed to experience similar symptoms to human beings. Therefore, it is important to remember that these new forms of trauma are not all clinically equivalent.

The expansion of trauma studies has been matched by a similar expansion in the remit of trauma theology. For example, Deanna Thompson's work addresses the ongoing trauma of living with incurable cancer. Her theology must grapple with a form of trauma that is endemic rather than catastrophic, and internal rather than external.[14] Meanwhile, Jones reflects on the growing number of contexts in which trauma is diagnosed. She says she has become increasingly aware of both collective and intergenerational traumas, including what she calls the "entrenched horrors of life in America," such as racism, white supremacy, misogyny, and violence against women.[15] She also highlights her growing sense of the importance of secondary trauma, where those who observe and accompany trauma survivors become traumatized themselves.[16] Likewise, Katie Cross and Karen O'Donnell are especially aware of the changing face of trauma theology. In their edited volume *Bearing Witness*, they explicitly focus on intersectional and chronic forms of trauma—including chapters on race, gender, sexuality, poverty, and health—as a contribution to what they call "the incomplete and ongoing work of trauma theology."[17] Trauma theologians are also becoming increasingly attentive to the ways in which ecclesial communities are sites of both trauma perpetration and post-traumatic response.[18]

Shelly Rambo's work serves as a good example of how theologies are being updated and reconceived to address newly recognized forms of trauma. In her first monograph, she focuses on the Triduum—the three days that span Christ's crucifixion and resurrection—and identifies the Spirit as a witness who remains in cases of traumatic suffering.[19] If approached alongside an individualistic atonement theology, there is a risk that her reading of the Triduum only speaks to individual human traumas. But, in her second monograph, Rambo observes a group of war veterans and their family members, which prompts a reflection on the communal dimensions of trauma.[20] Just as wider society needs to realize that the wounds of its war veterans play out as collective wounds within the community, so too we must understand Christ's wounds, not only as marks of an individual sacrifice, but also as interpersonally and politically potent.[21] More generally, Rambo notes how trauma studies has moved away from an examination of the individual human psyche to consider historical, institutional, global, and political forms of traumatic violence.[22] She highlights how:

> A growing awareness of trauma and its effects has also contributed to the expansive use of trauma to speak not only to the experience of particular persons and communities but also to a broader cultural ethos in which we live.[23]

Rambo picks up here on the way in which the terminology of trauma has found relevance in new domains. She conceives of trauma as a violence that is overwhelming in its impacts and therefore applicable in multiple contexts. Trauma theologians then have the task of tracking and responding to these new identifications of trauma.

That said, it is important to acknowledge that there are also critics of this expansion in trauma studies. The concern is that the apparent ubiquity of trauma diminishes its force and its relevance. This is primarily about the semantic relaxation in the definition of trauma, although there is some skepticism about the scientific discoveries too. Numerous scholars note that trauma has now undergone an "unbounded outward movement," "permeated every corner of our culture," turned into "a norm rather than an exception," and "become all-inclusive."[24] There is a risk that any gains in terms of comprehension come at the cost of a loss of precision.[25] If trauma permeates existence, then it becomes much less meaningful to diagnose or declare a specific instance of trauma. In Wulf Kansteiner's terms, "claims about the ubiquity of cultural trauma have quickly turned into unintended gestures of disrespect towards today's victims of extreme violence."[26] His concern is that the more we allow for phenomena such as collective trauma, the more we dilute the acute and concrete suffering of specific individuals. He continues, "the most severe abuses of the trauma concept currently occur in the abstract, metaphorical language of cultural criticism."[27] On Kansteiner's account, it is extremely important to distinguish between the irreducible horror of specific experiences of trauma and wider cultural representations of violence.

However, as Michael Rothberg pinpoints, some of the resistance to the widening of what counts as trauma stems from a competition model where traumatic histories "compete for attention within the public sphere."[28] On this understanding, the suffering of others must inevitably be diminished—thereby creating a "hierarchy of suffering"—in order to claim public attention for one's own cause.[29] But Rothberg's point, using Jewish history and the history of slavery as examples, is that, while each trauma is utterly unique, we need not subscribe to a zero-sum game. Rather, our memory can be "multidirectional," and survivors of different oppressions can learn through dialogue. It is shrinking the definition of trauma, not widening it, that risks diminishing the suffering of others. We must allow for the fact that a concept achieves its success through its heterogeneity rather than its purity.[30] Trauma theorists do not make progress by fighting for disciplinary vocabulary, but by seeking clarity in their own definitions and by being aware of how their use of a term both differs from and resonates with others. It is absolutely vital to remember that describing multiple different processes and events as traumatic is not to assert that they are phenomenologically or morally equivalent. Furthermore, allowing for an expanded understanding of what constitutes an instance of trauma is not the same as saying that trauma is completely ubiquitous. Certain characteristic features—such as the three diagnostic ruptures to communication, flesh, and time—do still have to be met for an ascription of trauma to be meaningful.

With these caveats in mind, it is nonetheless possible to see how the expanding concerns of trauma theology lead naturally to a consideration of ecological trauma. What Thompson, Jones, Cross, O'Donnell, and Rambo all demonstrate is that trauma theologians are already alert to new contexts in which their expertise is applicable. Trauma studies and trauma theology have always been disciplines that are open to extension in the light of new revelations about the occurrence of traumatic violence. Indeed, trauma theologians specifically aim to be attentive to forms of overwhelming oppression and injustice that have previously been silenced or ignored, since such examples of suffering are especially likely to have been traumatic. As Cross and O'Donnell state in *Bearing Witness*, they are very conscious of "what was missing" in their earlier work on trauma theology.[31] In the context of contemporary climate change and mass species extinction, multiple traumas appear to intersect and compound. The hurricanes, heatwaves, wildfires, rising sea levels, pollution, and biodiversity loss of the current ecological crisis are arguably some of the most extreme forms of suffering of the present age. In these circumstances, the Earth itself is subject to traumatic violence, and deserves to be considered by trauma theologians. To be clear, ascribing trauma to the Earth has more in common with the expansion in trauma terminology evidenced by collective trauma than it does with the proposed diagnoses of multispecies PTSD. As I argued in Chapter 1, it is challenging in Western cultures to assert an animist ontology as a basis for the consideration of ecological trauma. But we can, and should, speak of the traumatized Earth as part of a consciously anthropomorphic approach to fostering relationships. Moreover, the conceptual expansion of trauma theory is especially pertinent to ecological concerns. The acceptance that trauma arises from endemic as well as catastrophic violence means that, for example, the slow accumulation of carbon dioxide in the atmosphere can be considered traumatic for the Earth. And the recognition that trauma can emerge internally as well as externally allows for the fact that all possible ecological traumas emerge from within the Earth system, even if they entail anthropogenic causes. To reiterate, ecological trauma shares certain characteristic features with other forms of trauma, but it is also composed of different phenomena, different entities, and different processes. The expansion of trauma theology should allow for ecological trauma, but this new diagnosis should not be treated identically to individual human traumas.

Trauma and Interconnection

Hurricane Katrina made landfall in Louisiana on 29 August 2005. As it crashed into the Gulf Coast, it brought winds of 175 miles per hour and extensive flooding to more than three quarters of the city of New Orleans. Some 1,800 people were killed, hundreds of thousands were left homeless, and many more were forced to relocate. Estimates vary, but somewhere between a quarter and a third of those people living in areas directly affected by the hurricane subsequently met the diagnostic criteria for PTSD.[32] For years afterward, the damage to communities and infrastructure meant that death and morbidity rates were significantly elevated.[33] Both during and after the hurricane it was poor and Black communities that were the worst affected. Commentators also diagnosed the failure of the US government to respond adequately to the crisis as a form of cultural trauma. For example, Ron Eyerman documents "public articulations of collective pain and suffering" that were generated in response to the "tear in the social fabric" precipitated by the hurricane and its aftermath.[34] In short, Katrina was not just catastrophic, but deeply traumatic.

Hurricane Katrina is particularly important to the development of trauma theology because it forms a central case study in Rambo's *Spirit and Trauma*, one of the earliest monographs in the discipline. Rambo begins the book in Deacon Julius Lee's devastated backyard in the Lower Ninth Ward, the New Orleans district hit hardest by the hurricane. It is here that Rambo first articulates the persistence of traumatic suffering. As she writes, "Hurricane Katrina is not simply a singular event that took place in August 2005. It is an event that continues, that persists in the present. Trauma is what does not go away."[35] It is this persistence that prompts Rambo to start questioning the linearity of the Christian narrative of redemption, and to seek a theological witness that remains with trauma survivors.

One interpretation of Katrina sees the widespread trauma in the wake of the hurricane as a purely human phenomenon occasioned by a purely natural disaster. But Rambo's reading refuses such a straightforward separation. She is keen to highlight the way in which traumatic suffering arose from, and was exacerbated by, several interconnections between human and natural systems. She notes how "the trauma of Katrina cannot simply be limited to the violence of a natural disaster, but must be located within the broader and long-standing structures of oppression that existed long before the storm."[36] She also observes that "Hurricane Katrina sparked a deeper and more widespread storm of economic injustice, racism, and poverty."[37] Collective human traumas both contributed to, and arose from, the natural trauma of the hurricane. And there are further layers of entanglement between human and natural processes. Hurricane Betsy, in 1965, had prompted President Johnson to order the construction of levees in New Orleans.[38] But a series of shortcuts and human failings meant that these levees were insufficient to withstand Katrina's storm surge. Moreover, Katrina was not an isolated natural disaster, but part of a pattern of more frequent and more intense hurricanes resulting from anthropogenic climate change. New Orleans was retraumatized by Hurricane Isaac in 2012 and Hurricane Ida in 2021.

Rambo is also explicitly attuned to the ecological dimensions of Katrina's trauma. For instance, it is not just human beings who carry the imprint of traumatic suffering. As she writes:

> The streets and the land bear the marks of an overwhelming storm. Long after persons and communities experience trauma, the storm is still there. . . . There is something in the visible landscape of New Orleans that mirrors the internal processes of persons and communities living on in the aftermath of trauma.[39]

In other words, the landscape—indeed the very Earth itself—is marked by the trauma of the storm. It bears tangible ruptures just like the people and communities of New Orleans. As Lee recounts to Rambo, while standing in what remains of his backyard, "the storm is gone, but the 'after the storm' is always here."[40] This is true for human residents of New Orleans, and it is true for the nonhuman landscape too. There is a sense in which the Earth is also forever more "after the storm." What Rambo's analysis of Katrina implicitly reveals is the importance of engaging with the ecological dimensions of trauma. Many individuals suffered from PTSD, whole communities were collectively traumatized, and the nation underwent a form of cultural trauma, but Katrina also entailed traumatic suffering in the ecological realm. The hurricane arose, at least in part, as a traumatic consequence of global climate change, and went on to leave a traumatic mark on the landscape.

Rambo's sense of the ecological dimensions of trauma is also echoed by other scholars working in trauma studies. For instance, Erikson's examples of collective trauma—such as the 1972 Buffalo Creek flood in West Virginia—reveal similar ecological interconnections. The Buffalo Creek event resulted from the interplay of natural and human processes: the disaster was caused by both heavy rainfall and the failure of a coal slurry dam. In addition, there is something traumatic about the Buffalo Creek flood that exceeds the sum of the human traumas involved. What Erikson suggests is that "some form of the term 'trauma' is the most accurate way to describe not only the condition of the people one encounters in those scenes but the texture of the scenes themselves."[41] This awareness of trauma *in the scene itself* suggests that trauma studies, and trauma theology, must explicitly address the ecological realm.

Lauren Woolbright proposes that we call such scenarios "double trauma"— we traumatize the natural world, which then traumatizes us in return.[42] Natural disasters exacerbated by climate change are a particularly good example: we are responsible for the greenhouse gas emissions involved, but we are also killed, injured, and forced to migrate by events like Hurricane Katrina. It is only because our actions mesh with natural processes, and natural processes have an influence on us, that the true trauma of climate change emerges.[43] Yet, as I explained in Chapter 2, ecological thinkers increasingly reject any sharp distinction between nature and culture.[44] One of the distinct advantages of approaching ecological suffering through the lens of trauma is that there is no requirement to categorize the causes of such suffering as either natural or cultural before an appropriate response can be offered. What we draw from Rambo's discussion of Katrina, therefore, is that human and natural processes can never be fully disaggregated. Even Woolbright's "double trauma" might drive too sharp a wedge between the human and the natural realms. At every turn, human and natural processes are interconnected to the point that the nature/culture dichotomy breaks down.[45]

Elsewhere, Rambo is even more explicit about the unavoidable interconnections between human and ecological trauma. In the acknowledgments of *Spirit and Trauma* she says:

> I feel the fragility of the world more acutely than I did ten years ago. I view persons as more vulnerable in it, and the earth more wounded by our heavy footprints. I feel its weight. I have come to believe that we are connected in ways that we cannot account for and constituted by much that we do not know.[46]

While in a later publication she writes:

> Trauma studies confront us with the notion that we are constituted by the pain of others. This language of vulnerability also extends to the fragility of the earth and the planetary, as the realities of climate change and energy resourcing provide images of our wounded environment. We can speak about violence done to the earth and the consequences of exploiting resources differently when placing trauma within a broader picture of human and planetary interdependence.[47]

What Rambo identifies in these passages is that trauma theologians need to take account of "human and planetary interdependence." As she puts it, "we are connected in ways that we cannot account for" and even "constituted by the pain of others." Hurricane Katrina is just one example of how these interconnected traumas come to the fore. Moreover, Rambo is clear that trauma theologians need to engage specifically with the contemporary ecological crisis. She talks concretely about climate change, energy use, and the way that the fragile Earth is being "wounded by our heavy footprints." She also recognizes the likely inaccessibility of some of these ecological traumas—we are, she says, "constituted by much that we do not know." But much like Jones in the epigraph to this chapter, Rambo's conclusion is unambiguous: trauma theology holds the potential to speak to these ecological concerns. Trauma theologians must start addressing the violence that is perpetrated against the Earth.

Ecology in Existing Trauma Theologies

So far in this chapter I have observed how trauma theologies have naturally expanded to consider new cases of traumatic suffering, and explored the necessary interconnections between human and natural processes in many traumatic events. The implication of these observations is that a theology of ecological trauma is eminently plausible. To date, there have been no extended attempts to apply trauma theology in this way. There are, however, several trauma theologies that already include motifs from the natural world

Natural Metaphors

One way in which an ecological dimension already figures in many trauma theologies is through the authors' choice of metaphors. For example, Jones reaches for a comparison with a tsunami wave to try to communicate the severity of human trauma. She writes, "like the wave of a tsunami, they [traumatic events] drown you and disable your normal strategies for dealing with difficulties."[48] In the same vein, Jennifer Baldwin states, "the intensity of this ripple effect [of traumatic wounding] can range from that of a rough wave to a catastrophic tsunami."[49] Something about the world-changing reality of a tsunami seems appropriate to the life-altering impact of trauma.

Even more strikingly, O'Donnell includes an interlude in her monograph *Broken Bodies* that is simply entitled "Rupture." Here, she emphasises the profound similarities between a seismic rupture and a traumatic rupture: "Like an earthquake rolls through a landscape and radically alters the topographic features, so does trauma roll through lives, stories, memories and bodies, leaving them radically altered."[50] Elsewhere, she writes, "one might consider trauma to be an earthquake that shakes through a person—opening fissures, destroying structures, uprooting firmly held ideals—leaving behind it destruction, an unrecognisable landscape, and a deep sense of instability."[51] Later on in her interlude, O'Donnell draws on another natural metaphor to help describe the phenomenon of post-traumatic growth: "Like a forest awakening in the aftermath of a fire, or a trauma survivor stirring up a survivor's gift in the aftermath of trauma, some stories can only be told in the wake of the rupture."[52] In the same vein as the earthquake, the devastation of a forest fire is offered as an analogy for traumatic suffering.

In each of these cases—the tsunami, the earthquake, and the forest fire—the ecological comparison is intended to underscore the radical impact of human trauma. The logic is that human trauma becomes clearer by comparison with a self-evidently traumatic phenomenon from the natural world. But these same metaphors can easily be inverted: it is not just that ecological phenomena aid our understanding of human trauma, but also that the category of trauma can help to elucidate these ecological phenomena. If this is the case, then ecological traumas are already quietly present in all these trauma theologies.

Ecological Aspects of Trauma Hermeneutics

Theological trauma hermeneutics applies insights from trauma theory to the interpretation of scriptural texts. It reads violent imagery and fragmented narratives—including absences, gaps, and repetitions—as indicators of traumatization.[53] Rebecca Copeland specifically aims to expand trauma hermeneutics to cover ecological trauma, noting how trauma hermeneutics has hitherto "largely ignored the wider ecological systems that may have also been traumatized."[54] She defines ecological trauma as "the disordered state that results from severe stress or injury to a community of creatures."[55] Copeland proceeds to read the vision in Ezekiel 47:1–12 as an example of soil salinization and therefore as a trauma experienced by the land. She suggests that the native plants, the failed crops, the unstable water table, and the soil itself are all part of the traumatized community of creatures. The failure to grow food then results in the secondary human trauma of starvation. As she writes:

> Consideration of ecological trauma allows, and even requires, interpreters to expand their understandings of the traumatized community from which a text emerged to include the ecosystems upon which the human community depended.[56]

In other words, trauma hermeneutics must be sensitive to texts that arise not just from traumatized human communities, but also from traumatized ecologies. Copeland goes on to note that there is a tendency within the Hebrew Bible—and Ancient Near Eastern literature more broadly—to blame either the divine or unrelated human sins, such as sexual immorality, for the trauma experienced by the land.[57] But an ecological trauma hermeneutics focuses attention on a different sin: that of damaging agricultural practices.

What Copeland's plea highlights is that an ecological element often lies just under the surface of some of the work in trauma hermeneutics. For example, several trauma theologians turn to laments from the Hebrew Bible for a vocabulary with which to articulate traumatic suffering. Biblical examples of lament give survivors permission to be angry at, and protest against, God for traumatic situations of suffering and injustice.[58] The lament psalms provide an especially helpful resource because they often contain disclosures of traumatic suffering.[59] In Psalm 77 we hear of the traumatic rupture of communication, when the psalmist writes: "I am so troubled that I cannot speak."[60] Likewise, Psalm 38 does not hold back: "I am utterly spent and crushed; I groan because of the tumult of my heart."[61] As Megan Warner notes, the lament psalms "address the need for witnessing."[62] Although many psalms move from lament to praise, and therefore move rather too seamlessly from trauma to healing, there are other psalms that move in the opposite direction.[63] Psalm 88 is especially notable for never resolving the lament, ending in "a sense of utter isolation, with darkness as the speaker's only companion."[64] The psalmist simply stands with the wounded.

Yet the psalms, and other examples of lament in the Hebrew Bible, are also central for many ecotheologians. Indeed, it is claimed that the psalms could play an important role in ecological conversion since: their many references to creation provide models for relating to the natural world; they were composed in a premodern culture that was more in tune with the Earth; and their regular use and performance can help to change minds and shape behaviors.[65] Furthermore, as I identified in Chapter 1, the psalms and the prophetic literature are replete with examples of the devastated land and the mourning Earth, and these scriptural anthropomorphisms also have the power to shape how we relate to the natural world.[66] Isaiah recounts how "the earth is utterly broken, the earth is torn asunder, [and] the earth is violently shaken."[67] This reads like another disclosure of trauma, only this time it is the trauma of the Earth that is being reported. Just as other laments enable an expression of human trauma, this example from Isaiah provides vocabulary for an articulation of *ecological* trauma. In the psalms and the prophets, the planet both laments and is to be lamented.

All of this suggests that there exists a largely untapped ecological dimension to the scriptural lamentations employed by trauma theologians, and Copeland's ecological trauma hermeneutics specifically encourages us to bring these readings to the fore. There is an existing ecological thread running through this branch of trauma theology.

Ecological Elements at the Crucifixion

A third focus within trauma theology where ecological elements are already at play is the crucifixion of Christ. For several trauma theologians, the crucifixion is deemed to be a paradigmatic example of traumatic rupture. According to Jones, it is hard to envisage a "more traumatic event than the torture and execution of this man Jesus."[68] Or, as Hilary Scarsella observes, "without the traumatic event of Jesus's crucifixion . . . Christian theology as we know it would not have come into being."[69] The epitome of this trauma is encapsulated in Christ's cry of dereliction from the cross: "My God, my God, why have you forsaken me?"[70] Christ's cross has therefore become a symbol of solidarity for many subsequent Christians experiencing traumatic pain.[71] Yet the crucifixion constitutes an archetype of traumatic suffering, not just because of the pain that Christ himself experienced, but also because of the effect it had on Christ's followers, and the terror it caused in the wider context of imperial rule.

Furthermore, there is also an ecological dimension to the crucifixion. As Matthew's gospel records, "darkness came over the whole land," and then, at the moment of Christ's death, "the earth shook, and the rocks were split."[72] The crucifixion is accompanied by an earthquake. Perhaps the most common interpretation of the sudden twilight and the seismic shaking is that they are apocalyptic motifs, signs of the eschatological future to come. But they also point to the possibility that the whole of creation is drawn into this intense moment of trauma. Some scholars read the crucifixion earthquake as God speaking through nature; others see it as nature responding in fear to the sovereignty and judgment of God; but a third possibility would be to interpret the earthquake in terms of the voice and activity of the Earth itself. As Norman Habel maintains, "Earth is a subject capable of raising its voice in celebration and against injustice."[73]

In her volume for the Earth Bible commentary series, Elaine Wainwright develops this reading of the crucifixion earthquake in terms of the voice of the Earth. She notes how the "Earth itself speaks" through Matthew's crucifixion earthquake.[74] If Christ's cry from the cross encapsulates the Christological trauma, then it is conceivable to think that the crucifixion earthquake encapsulates the trauma of the Earth. The two traumas are chiastically intertwined. As Alan Cadwallader writes, "God joins earth in the agony."[75] Or, as Wainwright puts it, "it is as if Earth mourns the profound and absolute abjection of Jesus," while Christ's cry of godforsakenness evokes "the cry of innocent and suffering Earth."[76] The earthquake's function here is to represent all the traumatic suffering in the natural world. On this understanding, it becomes impossible to separate the ecological dimension from the events of the crucifixion trauma. The Earth's own suffering is already expressed in what trauma theologians frequently regard as the original and central trauma of Christian theology. Ecological suffering is embedded at the heart of the crucifixion narrative.

* * *

Ultimately, ecology deserves greater consideration in trauma theology because, quite simply, it is already tacitly present: in the metaphors from the natural world, in the laments of the Hebrew Bible, and even at Christ's crucifixion. Some of these ecological dimensions are more explicit than others, but their latent presence further highlights the potential of a theology of ecological trauma.

Conclusion

I began this chapter with a quotation from Jones. We are seeing, she says, trauma that is "perpetrated against the earth itself," and she perceives the ecological promise of her own work in trauma theology.[77] I then provided three arguments as to why future work in trauma theology ought, as one of its aims, to be engaging with the ecological realm. First, trauma theology has always been revising and updating its focus as new survivors of overwhelming and extreme violence are uncovered, and new entities are diagnosed as traumatized. The advent of ecological forms of trauma is just the latest iteration of this trend, and trauma theologians should be seeking to respond. Second, Rambo's writing about Hurricane Katrina illustrates how trauma theologians are already conscious of the interconnected character of many traumatic events. Trauma may be diagnosed, not just in people and societies and cultures, but also in devastated landscapes and environments—in the very scene itself. The nature/culture dichotomy can unhelpfully occlude recognition of ecological trauma, and this is an issue that trauma theologians should be seeking to address. Finally, existing work in trauma theology already admits to an ecological aspect. These vignettes highlight possible ways forward for a theology of ecological trauma: in terms of metaphor, in terms of lamentation and witnessing, and in terms of Christ's relationship to the Earth. These are all themes that will recur in the second half of this book.

In Chapter 2, I argued for the inclusion of trauma in ecotheology, and now, in this chapter, I have argued for the inclusion of ecology in trauma theology. Yet this double novelty lands at a single intersection. There is something to be explored at the boundary between these theological subdisciplines. The introduction of trauma to ecotheology and the expansion of trauma theology to include ecology both involve applying the methods and motifs of trauma theology to questions of ecological suffering. As the examples in this chapter have shown, there are undoubtedly multiple ways of carrying this out. But what both chapters have ultimately sought to demonstrate is one and the same need—for *a theology of ecological trauma*.

4
The Rupture of Communication: Christ's Witness to a Wounded World

> Staying with the trouble requires learning to be truly present, not as a vanishing pivot between awful or edenic pasts and apocalyptic or salvific futures, but as mortal critters entwined in myriad unfinished configurations of places, times, matters, meanings.[1]
> —DONNA HARAWAY, *Staying with the Trouble*

Donna Haraway is well known for her notion of "staying with the trouble." Not diagnosing or analyzing the trouble. Not resolving or dismissing the trouble. But simply staying with the trouble. What is especially striking in the epigraph above is Haraway's disinterest in both origins and destinies. On the one hand, she brackets the causes of the current trouble: the fall from Eden has no immediate bearing on what should be done now. On the other hand, future-facing eschatologies are similarly off-limits: no promise of salvation to come can do justice to the messy configurations of the moment.[2] As she goes on to state, "I am not interested in reconciliation or restoration, but I am deeply committed to the more modest possibilities of partial recuperation and getting on together. Call that staying with the trouble."[3]

When Haraway speaks of trouble, she has in mind various aspects of the contemporary socio-ecological crisis. What she proposes is that talk of future ecological restoration be put on one side in favor of a deep engagement with present realities. In forestalling a focus on the future, she seeks to undercut both techno-optimism and resigned fatalism. Hope for some sort of external savior, whether technological or theological, is of little importance to Haraway if we are not willing to contribute to the hard graft of staying with the trouble in the present. But giving up is not an option either; to despair is to abandon the trouble just as much as any facile hope. Instead, we must stay with the gritty realities of "unfinished configurations of places, times, matters, meanings," resolutely confronting the present. This means that the task of staying with the trouble is humble, fragmentary, and difficult. It can involve "partial recuperation" and "germs of partial healing," but it is also about "getting on together" and "learning to be truly present" in the here and now.[4] This is how to live with the trouble.

In this chapter, I propose a version of staying with the trouble as a plausible response to the recognition of ecological trauma. In the Introduction, I followed Shelly Rambo and Karen O'Donnell in characterizing trauma in terms of three ruptures—to communication, to flesh, and to time.[5] Over the course of the previous chapters, I have suggested that theologians need to specifically engage with *ecological* forms of trauma. Throughout, my aim has been to employ the "lens of trauma" as a heuristic tool for reconsidering theological approaches to ecological suffering.[6] The task in the remaining three chapters is to address the traumatic ruptures of communication, flesh, and time as they pertain to the ecological realm and, in each case, to offer a theological response.

Here, I begin with the rupture of communication. When describing human trauma, Rambo and O'Donnell refer to a rupture in cognition or language, or even just a rupture of "word."[7] But the broader category of communication allows for noncognitive, nonlinguistic, and nonverbal processes, which are often important in nonhuman contexts. In the case of ecological trauma, the phrase "rupture of communication" refers primarily to a breakdown in communication between human beings and other elements of the natural world. As perpetrators of the sixth great mass extinction, we seem to be immune to the pleas of disappearing species. As the guilty culprits of anthropogenic climate change, we are failing to properly understand the cries of an inexorably warming world. Contemporary ecological suffering is occurring on vast scales and yet we appear not to hear. To borrow the language of trauma theorist Dori Laub, it is as if there has been a "collapse of witnessing."[8]

Addressing Laub's collapse of witnessing in the ecological realm is going to require bearing witness to instances of ecological trauma—and this is what looks a lot like Haraway's notion of staying with the trouble.[9] Bearing witness to traumatic ecological suffering involves wrestling with the silence, unearthing the pain that has been suffered, and sharing it with others in the hope that this pain will be recognized and dignified. This is about living alongside and remaining present. It is by bearing witness that ruptured lines of communication might tentatively begin to be restored. In what follows, I turn to Rambo for a theological account of witnessing that remains sensitive to trauma survivors. I go on to suggest that, for Christians, Christ can serve as a witness to ecological trauma, and I compare this model of Christic witnessing with other theological responses to ecological suffering. Importantly, Christ is not just a passive observer, but an active witness to the wounds of the world.

Witnessing Trauma

In her book *Spirit and Trauma*, Rambo problematizes two common theological understandings of witnessing. Christianity, she suggests, has often tended to see witnessing as either proclamation or imitation: a list of facts about Jesus, or an attempt to emulate his life and his sacrifice. But each of these understandings is limited in relation to trauma.

According to Rambo, the proclamation model of witnessing imagines an onlooker, bystander, or spectator who gives verbal testimony as if in a court of law.[10] The purpose is to impart information and, in the case of Christian witnessing, to share beliefs about Jesus with the aim of bringing about conversion in the recipient. But the problem with this practice of proclamation is that it assumes a degree of mastery, a sense of understanding and control, that is often completely lacking in cases of trauma. Trauma survivors, and those who encounter traumatic phenomena, rarely have the ability to explain exactly what has taken place. This is because trauma's rupture of communication destabilizes not just our own certainties, but also the very reliability of language to convey what we might think we understand. Witnessing, says Rambo, "involves not only a gathering around an enigma but also a process of handing over and receiving an experience that continually refuses straightforward communication."[11] Bearing witness to trauma is not nearly as straightforward as mere proclamation.

In Rambo's imitation model, the witness attempts to copy and participate in the life of Christ—his love, his service, and his sacrifice—by patterning their own life after the embodied practices of Jesus.[12] This somatic mimesis may even include risking one's life for one's faith; if Christ suffered and died for the truth, then we must be willing to do the same. Yet this model of witnessing also has its problems. Imitation can suppress difference and encourage sameness.[13] It can fail to recognize the other's trauma as truly other and so it can marginalize the other's suffering in the act of imitation. Even more alarming, an imitative model of witnessing can be co-opted to recommend self-sacrifice and glorify martyrdom.[14] Indeed, many feminist and womanist theologians highlight the dangers of fixating on self-sacrifice.[15] At its worst, a recommendation to imitate self-sacrifice can encourage women to accept abuse or serve as surrogates. Violence is promoted rather than curtailed. Christian witnessing, by contrast, ought to be about something much less self-destructive and much more open.

Translated into the ecological realm, conceptions of witnessing as proclamation or imitation are likewise rather limited. We have, for decades, been proclaiming facts about climate change and species extinction, but with little discernible impact. This sort of witnessing is restricted in what it can achieve. This approach also assumes that we have fully mastered the information about ecological breakdown that is being reported when there is in fact something inaccessible and unassimilable about the current ecological crisis.[16] Similarly, attempting to imitate examples of ecological suffering in our own lives makes little sense in this context. Human beings form part of the natural world and so self-destruction contributes to, rather than prevents, planetary destruction—a point that is missed in the misanthropic suggestions of some of the more extreme environmentalists.

Instead, for Rambo, witnessing should be about "remaining" with the reality of suffering—something that cannot be fully comprehended—even when everyone else is impatient to move on.[17] Witnessing, she says, "describes a way of being oriented to what remains, to the suffering that does not go away."[18] It involves staying with trauma survivors in the present and being brutally honest about what they are experiencing. There is no heroic guarantee of salvation or even any promise of future justice. A witness is simply "what endures, survives, or is left over after everything else falls away."[19] But this is not an easy thing to do; avoiding or ignoring the trauma is a much more common response. Rambo draws here on her interpretation of the farewell discourse in John's gospel, where Jesus says goodbye to his disciples on the night before his crucifixion. Jesus's instruction to his disciples to *menein* is usually translated as "abide," but Rambo reads this as a direction to "remain," to wait and to endure, during the upcoming crucifixion and beyond.[20] To use Haraway's idiom, Rambo is insisting on staying with the trouble. Such witnessing evidently does not take the form of either confident proclamation or self-sacrificial imitation, but focuses instead on the more mundane yet ultimately more demanding task of remaining with the trauma. The purpose of such witnessing is to bring trauma to light, to dignify past suffering with recognition, and—hopefully—to reduce future violence. But what is most crucial is the movement to remain.

Rambo's proposal of witnessing as remaining is especially appropriate to ecological trauma because it resonates with recommendations from other disciplines and literatures. For example, in *Witnessing: Beyond Recognition*, Kelly Oliver makes a similar argument about witnessing as something other than the recounting of facts or the copying of postures. Witnessing, says Oliver, goes beyond recognition, "testifying to that which cannot be seen," and straining for the unfamiliar at the heart of traumatic suffering.[21] What is needed is not comprehension, but encounter. Moreover, for Oliver, subjectivity is constructed through the act of witnessing itself.[22] Processes of oppression and traumatization frequently involve objectification of the survivor, whereas witnessing restores subjectivity by enabling the survivor to speak. This has radical consequences in the case of ecological trauma: witnessing the planet's trauma begins to confer subjectivity on the Earth itself. Admittedly, the argument is somewhat circular here: we can only recognize ecological trauma by anthropomorphizing nature, and then, when we bear witness to that trauma, we begin to construct the nonhuman subjectivity that was required for the ascription of trauma in the first place. However, the critical point for the ecological trauma theologian is not to secure the truth of certain ontological statements, but to recommend a practice that improves humanity's relationship to the rest of the natural world in the midst of ecological suffering. Witnessing as remaining goes beyond recognition because it is present to that which cannot be comprehended or articulated. In the process, it spotlights the subjectivity of the survivor.

Likewise, in her book on climate education, *Learning to Live with Climate Change*, Blanche Verlie specifically advocates witnessing, alongside encountering and storying, as a suitably affective and embodied response to the "deeply traumatic" phenomenon of climate change.[23] As Verlie recounts, witnessing is a practice that enables us to "acknowledge, articulate, inhabit, enact and respond" to climate change while simultaneously relinquishing our desire for control.[24] Verlie is also clear that bearing witness involves standing *with* others—"wit(h)nessing"—rather than keeping a supposedly objective distance.[25] Climate science, says Verlie, typically aims to produce a "testimony of planetary trauma" in the form of "unemotional models, predictions, and statistics."[26] Consequently, climate science sometimes feels too abstract and too disengaged from the realities of ecological trauma. But climate witnessing—remaining with the testimonies that emerge from tree rings, ice cores, dry riverbeds, saline drinking water, mass animal death, bleached coral reefs, and so on—is an affective labor that forms feeling relationships. It does not necessarily solve climate change, or even improve our knowledge of climate change, but it does help us to live with climate change. For Verlie, too, this is about staying with the trouble.[27]

Christ as Witness

So, where might a Christian theologian turn for a suitable witness? Is there a good theological example of witnessing as remaining? And can this witness address the rupture of communication in the ecological realm?

Within her own project, Rambo turns to the Holy Spirit as a witness who remains through the trauma of the Triduum.[28] Rambo is interested in Holy Saturday, a focus she draws from the writings and visions of Hans Urs von Balthasar and Adrienne von Speyr. For Rambo, the theological question is: who or what bears witness to this trauma? "Despite the christological renderings," says Rambo, "I suggest that this witness . . . is better understood pneumatologically."[29] Rambo is reticent about Christology because she worries that Balthasar and Speyr are too quick to read Holy Saturday in terms of an imitative model of self-sacrifice.[30] If Christ is the principal witness, then those who seek to follow Christ are called "to enter into the sufferings of Christ in an extreme way."[31] This entails two major problems. First, only an elite few, such as Speyr herself, can enter this "privileged mystical space" on Holy Saturday. Everyone else must "follow at a distance."[32] Second, this results in an understanding of Christian witness as martyrdom.[33] On such an account, we are called to take up our cross and obediently submit ourselves to violence, with all the problems that this entails. So, Rambo turns to pneumatology for an alternative. For her, the Holy Spirit, or "middle Spirit," is "the invisible means of transport" between Good Friday and Easter Sunday, bearing witness to the permeable boundaries between life and death.[34] The language of the middle serves to highlight the placement of this witness: embedded in present realities, but with only partial access to either the past or the future. This middle ground is the "between"—the "hinge between tragedy and triumph," as Rambo puts it—where the witness remains with irresolution.[35]

Yet the witness of Rambo's middle Spirit faces a problem. Trauma is an inherently somatic phenomenon, whereas Rambo's middle Spirit is disembodied and disengaged from the flesh. It is therefore hard to envisage how the Spirit bears witness to the trauma of the crucifixion inflicted on Christ's flesh. As Preston Hill argues, God requires a body if God is to bear witness to the crucifixion trauma and the post-traumatic stress of Holy Saturday.[36] Moreover, the wider concern here is to uncover a theological witness to *ecological* trauma. The processes involved, from deforestation scars to calving glaciers, are material and fleshly, so an embodied witness is especially important. Rambo's trauma theology is disconnected from material reality just at the point that it is needed most.[37]

Indeed, Rambo herself is aware of the importance of a somatic witness, and she points toward two possible solutions. First, she reads the "handing over" of Christ's spirit as a passing of the baton to the disciples gathered at the foot of the cross.[38] In this exchange, Rambo hints that the disciples are now the embodied witnesses of the crucifixion trauma: "the focus of the text is on the witnesses' bodies. . . . They physically testify to the absence that they are witnessing."[39] Hill elaborates on Rambo's argument: "God experienced the aftermath of Christ's trauma through the bodies of these disciples," he says.[40] This seems plausible inasmuch as the disciples could easily be the bearers of secondary or vicarious trauma, but it is less clear how the disciples can provide any fleshly continuity with Christ's resurrection body.[41] Rambo's second strategy is to emphasize the material characteristics of her middle Spirit. She writes, "the Spirit is not immaterial. Traditional understandings of Spirit as a disembodied entity have been 'asphyxiating,' cutting off the primal breath of Spirit."[42] Her hope is that the breathy connotations of the Spirit are sufficient to ground her pneumatology in the material realm. But, despite Rambo's best efforts, Christians still tend to envisage the Spirit in disembodied terms.[43]

It therefore seems, at least for a theology of ecological trauma, that a Christological witness is preferable to a pneumatological witness.[44] This is partly because of the unambiguous materiality of Christ's flesh: Christ maintains an embodied connection both to the trauma of crucifixion and to the fleshly nature of the world. But a Christic witness is also preferable because, as Rambo goes on to discuss in her subsequent work, *Resurrecting Wounds*, it is the scars on Christ's resurrection body that continue to bear witness to the trauma that has occurred. As Hill concludes:

> Since the body keeps the score of trauma, and this body belongs to a single human individual who experiences both the original traumatic event and the posttraumatic stress, it seems to make more sense to say . . . that this individual human body was that of the crucified Christ.[45]

In fact, Rambo tacitly acknowledges that Christ might be reinstated as the central Christian witness to trauma, albeit in a nuanced and chastened form, in that her pneumatological focus in *Spirit and Trauma* gives way to a Christological focus in *Resurrecting Wounds*. However, if we are to avoid repeating Rambo's well-founded concerns about a Christology that glorifies martyrdom, then two provisos must be borne in mind. First, this Christic witness must be built around a desire to remain with the suffering, rather than to imitate the violence. Second, significant care must be taken when translating this Christic witness into a model for Christians to emulate. A Christic witness must not end up recommending self-sacrifice.

Furthermore, this model of Christic witnessing consciously reorientates some of the more common understandings of Christian witnessing. Much of the discussion about Christian witnessing is interested in how humanity can bear witness to the Christ event, or in how Christ bears witness to God.[46] These are perfectly reasonable enquiries, but the suggestion here is that another form of witnessing is also occurring: that in Christ we find a theological witness to the world and its traumatic suffering.[47] This is not an unprecedented proposal. For example, liberation theologians like Jon Sobrino write that "Christ is the definitive witness," whose "power of witness" consists in "faithfulness to the cross."[48] This understanding of Christ's witness as consisting in faithfulness is a powerful echo of Rambo's insistence that witnessing is about remaining. Sobrino also says that God's suffering in Christ is a "silent witness" to "this world's victims," directly signaling the role that Christ could play in response to ecological trauma.[49] Christians can therefore think about Christ as someone who stays with the trouble.

Christ as Ecological Witness

So far in this chapter I have been proposing Christic witnessing as a theological response to trauma and the traumatic rupture of communication. But how might Christ's witness extend to the rest of the natural world and its sufferings? There are several possibilities, yet there is also a real danger here. Expanding the focus to the ecological realm may rightly prompt a concentration on the cosmic dimensions of Christ. But enlarging the field of view in this way risks distancing us from specific instances of traumatic suffering. An overly holistic Christology could result in insufficient attention being paid to the trauma of individual creatures.[50] Stating Christ's universal compassion for all creatures is not much help unless some form of historical justice-making is also enacted for individual trauma survivors. One possible counterbalance to this concern, proposed by Elizabeth Johnson, is to refocus attention on the particularities of the historical Jesus.[51] Throughout his ministry Jesus engaged with the plight of specific individuals. Scriptural motifs from the life of Jesus are therefore an important starting point for developing a Christic witness to ecological trauma.

"Look at the birds . . . consider the lilies," says Jesus in the Sermon on the Mount.[52] Many ecotheologians appeal to such lines as a general indication of the importance of the material world.[53] But there is also something more specific and more active about Jesus's plea. Here, at the heart of his preaching, he entreats the listener to pause and consider the ecological realm. His followers are not asked to make use of the lilies, or even to protect them, but simply to "consider" them, to honor them with undivided attention.[54] Furthermore, the thrust of this passage from Matthew's gospel is to provide reassurance about divine care no matter what trials or tribulations the birds and the lilies may suffer. There is a pledge to remain through thick and thin. What is especially striking is that Jesus's invocation to "consider the lilies" orientates the listener toward the importance of the present moment. The passage ends with the lines, "so do not worry about tomorrow, for tomorrow will bring worries of its own. Today's trouble is enough for today."[55] Just as Haraway's staying with the trouble guards against unhelpful speculations about the future, Jesus's words refuse to move on from "today's trouble." This is an act of witnessing that includes two of the essential components of any witness to ecological trauma: an active foregrounding of the nonhuman and a commitment to the present trouble—and Jesus is fulfilling the role of ecological witness.

The same dynamics are at play in the brief parable of the sparrows later in the same gospel. "Are not two sparrows sold for a penny?" asks Jesus, "yet not one of them will fall to the ground unperceived by your Father."[56] Denis Edwards reads these words as a reassurance that "Jesus looks on sparrows and other creatures with compassionate and loving eyes."[57] More specifically, given the suffering and death endemic to the process of evolution, Edwards argues that some form of eschatological redemption will be offered to "every sparrow that falls to the ground." Yet Edwards's speculations about salvation, while perfectly theologically legitimate, move beyond the immediate context of this passage. Rather, the emphasis here is on the fact that no sparrow's suffering will go "unperceived" by God. As Jesus states a few lines earlier, "for nothing is covered up that will not be uncovered."[58] Again, what is offered is a model of theological witnessing. There is a raw honesty about the suffering of sparrows and a promise that no suffering will be occluded. But within the passage itself there is no mention of either past causes or future hopes; no subsequent redemption is offered as an excuse for present suffering.

The broader point is that there are numerous aspects of Jesus's ministry that can be understood as acts of ecological witness. For example, Jesus has often been understood as both an ecological steward and a gardener.[59] Moreover, many commentators note how Jesus's parables are full of ecological allusions.[60] As Johnson puts it, they are "salted with references to seeds and harvests, wheat and weeds, vineyards and fruit trees, rain and sunsets, sheep and nesting birds."[61] Meanwhile, Edwards remarks on how "the beauty of wildflowers, the growth of trees from tiny seeds, crops of grain . . . the birds of the air, foxes and their lairs, rain falling," and so on, all feature in Jesus's parabolic teaching.[62] On one level, such parables can be interpreted as mere illustrations with no wider ecological importance. But, on another level, they seem to indicate that "the sense of the divineness of the natural order is the majority premise of all the parables."[63] Hence, these parabolic references can also be read as direct invitations into a deep contemplation of earthly phenomena and their significance. The very act of telling these parables is itself a witness to the Earth and its sufferings. Again, it is Jesus who is fulfilling the task of bearing ecological witness.

Cruciform Witnessing

Yet arguably the most obvious way in which Christ bears witness to ecological trauma is at the crucifixion. The link between Christ's wounds and planetary suffering is potentially profound. But this approach is not without its dangers. We must remain mindful of an excessive focus on the violence of the cross, even as we seek to tease out its relevance for a wounded world.

The centerpiece of Serene Jones's collection *Trauma and Grace* comprises three short essays on the cross: the alluring cross, the mirrored cross, and the unending cross. These chapters explore the role that the cross can play for survivors of trauma, including an awareness of the appeal of the cross, the solidarity offered by the cross, and the persistence of the cross. Jones is prompted in her thinking by the words of a trauma survivor: "This cross story . . . it's the only part of this Christian thing I like. I get it. And it's like he gets me. He knows."[64] The cross is meaningful to this trauma survivor because it echoes their own experience, enabling them to speak in terms of God's solidarity with their own distress. Jones calls this a Christology of mirroring: the power of the cross lies in its ability to reflect our stories of suffering back to us.[65] In relation to another survivor she writes:

> Cruciformed, he embodies the fractured, tortured shape of her traumatic existence . . . there is an identification of being that makes communication possible. . . . In bearing her story, he brings voice to the horror she herself cannot speak. He is, in other words, her embodied testimony. . . . He witnesses her; he receives her unraveled testimony-of-a-life as an offering of truth, and in that exchange, he articulates her unspoken history, her invisibility made visible in his eyes. . . . He assumes her reality, speaks the unspeakable in his own loss of speech, and then returns all of this to her as he witnesses to what she believed would be forever unknown.[66]

Jones's understanding of the crucified Christ as a witness to human trauma is especially apposite because it begins to address the rupture of communication that is inherent in traumatic phenomena. Christ "brings voice to the horror," revealing what would otherwise have been "forever unknown," primarily by providing an "embodied testimony." It is Christ's own embodied "loss of speech" that bears witness to that which is unspeakable.[67] A channel of communication begins to be reestablished for this trauma survivor in a fragmentary and fleshly form. The importance of the enfleshed character of this Christic witness is undeniable; without a body, Christ would not be able to provide the same testimony for this survivor. Furthermore, in the third of Jones's essays on the cross she emphasizes its "unending" nature.[68] This cruciform witness is not curtailed by any chronological conclusions but persists with the suffering. This is a witness who remains.

Yet the ecological dimensions of this Christology are not the focus of Jones's concern. She makes no mention of the nonhuman creation as she describes the work of Christ and the mirrored cross. As it stands, this cruciform witness remains highly anthropocentric.

For Christ to bear cruciform witness to the wounds of the world requires expanding Christ's solidarity to the whole of the nonhuman realm.[69] According to Johnson, the divine solidarity of the cross must be extended "into the groan of suffering and the silence of death of all creation."[70] The cruciform testimony about which Jones writes so poignantly needs to embody not just the tortured shape of human traumatic existence, but that of ecological trauma too. To rephrase the quotation from Jones, we might say that:

> Cruciformed, he embodies the fractured, tortured shape of *planet Earth* . . . there is an identification of being that makes communication possible. . . . In bearing *the planet's* story, he brings voice to the horror that *the planet* cannot speak. He is, in other words, *the planet's* embodied testimony.[71]

Again, the Christic witness reestablishes a form of communication, this time between the suffering Earth and ourselves. The traumatic rupture of communication is not easily healed, but Christ's embodied testimony conveys, albeit incompletely, some of the pain of ecological trauma to humanity. Christ gives a kind of "voice" to the apparently "voiceless" creation.

Furthermore, as I made clear in Chapter 3, several ecological elements are already at play in the narrated events of the crucifixion. The gospel record of darkness over the whole land, followed by an earthquake at the moment of crucifixion, demonstrates a link between Christ's trauma and ecological suffering. It is almost as if these ecological elements are inserted into the text as cinematic flashbacks to remind the reader of the scope of Christ's witness. To quote Elaine Wainwright again, Christ's cry of godforsakenness evokes "the cry of innocent and suffering Earth."[72] Note too how Christ's witness redirects humanity to the testimony of creation itself, reminding us of the possibility of a direct encounter with the Earth.[73] In this way, Christ's cry from the cross bears witness to the traumas of the Earth. Johnson highlights the same possibility when she cites a dramatic quotation from Arlen Gray:

> In his final death scream Jesus gathered up all of the earth's suffering throughout all time, bound it up and presented it before the heavenly throne, not in reams of words but in a sacred package encompassing all the sorrows, the sufferings, the lost dreams of all creation, all peoples, all times, all conditions.[74]

This Christic witness—which is aimed at God, as well as at us—is to "all of the earth's suffering throughout all time." Christ's communication here is also viscerally nonlinguistic. As Gray puts it, the cry is "not in reams of words" but takes the form of a "death scream." Hence, Christ's godforsaken scream, together with his embodied testimony, enable him to bear cruciform witness to the traumas of a wounded world.

Comparing Christic Witnessing to Other Theological Responses

At this juncture, it is important to explain how my proposal of cruciform Christic witnessing overlaps with, but also differs from, other similar responses to ecological suffering from within the existing literature. Many of the ideas below are not specifically proposed as responses to *traumatic* ecological suffering. Furthermore, several of these ideas are principally concerned with ecological soteriology, that is, with the issue of how creation might ultimately be transformed and redeemed. For the ecological trauma theologian, securing ultimate salvation is secondary to the more immediate problem of providing a theological response in the moment. Nevertheless, the following proposals represent a range of reactions to ecological suffering that could theoretically be employed by a theology of ecological trauma.

Cruciform Creation

Holmes Rolston derives his notion of the cruciform creation from a consideration of the evolutionary process. Evolution by natural selection produces new life by means of struggle, destruction, and death.[75] This leads Rolston to propose that the whole of creation can be considered cruciform, since it follows a pattern in which new life is brought about through the process of suffering, just as Christ's resurrection is wrought via his crucifixion. "The story we have from Darwinian natural history," he says, "echoes classical religious themes of death and regeneration."[76] For Rolston, this pattern of death and new life is not just a superficial quirk of evolution, but a religious principle that is sewn into the very fabric of creation, prefiguring Christ's own crucifixion. As he writes, "since the beginning, the myriad creatures have been giving up their lives as a ransom for many. . . . Jesus is not the exception to the natural order, but a chief exemplification of it."[77] It is important to emphasize that Rolston sees these creaturely sacrifices, not as an indication of the fallen world, but as part of the structure of creation itself. As he puts it, "the way of nature is the way of the cross; *via naturae est via crucis*."[78]

Rolston's approach is well suited to the methodology of the trauma theologian in that he resists any premature talk of redemption. The attribution of cruciformity to creation also suggests a connection between Christ and the Earth. It becomes relatively straightforward to imagine that Christ suffers in solidarity with the creation or is even present within the suffering creation. Such solidarity or presence could form the basis of a Christological response to ecological trauma.

However, there are problems with this proposal. For Rolston, the cruciform state of creation is not so much something to be protested and lamented as something to be celebrated for its creativity and productivity. In Rolston's view, "creativity is through conflict" and "success is achieved by sacrifice."[79] Rolston's cruciform creation becomes not just descriptive of the way things are, but prescriptive of how they ought to be. Yet if suffering, conflict, and sacrifice are rendered necessary and productive, then there is no stimulus to object to ecological violence. As Celia Deane-Drummond writes, "the notion of cruciform nature . . . seems to subtly endorse such suffering rather than giving the moral imperative to seek its amelioration."[80] Moreover, ecological trauma is not teleological in the way that Rolston conceives of the cruciform creation. Traumatic ecological suffering is not a case of, as Rolston puts it, creation "suffering through to something higher."[81] An additional concern with the idea of the cruciform creation is that it fails to address the suffering of individual creatures.[82] Rolston argues for a systemic perspective in which particular cases of ecological suffering are seen as less relevant than the good of the evolutionary process as a whole.[83] The specificity of individual ecological traumas can easily get lost.

These issues with the cruciform creation also raise a significant concern for models of Christic witnessing. The idea that Christ shows solidarity with, or is present within, the suffering entailed by ecological trauma is undoubtedly helpful for a theological response. But it is important to ascertain whether Christ mirrors the cruciform creation, or the creation mirrors the cruciform Christ. In fact, it turns out to be vital for models of Christic witnessing that Christ's witness to the suffering creation occurs as a subsequent step, rather than a preemptive recommendation. Contra Rolston, the logical ordering must be that creation is suffering, and then Christ bears witness to it at the crucifixion. This is not necessarily a temporal sequence—plenty of ecological suffering has occurred since the Christ event—but it does indicate a flow, or movement, *from* the suffering creation *to* the Christic witness. If the situation were reversed, and creation were to be mirroring Christ, then the suffering of creation would become necessary for creation to adequately emulate Christ. If cruciformity is creation's *telos*, then traumatic suffering becomes an obligation. As Edwards puts it, the cross "is not to be seen as some kind of necessary outcome of creation, or as a principle behind creation."[84] Insisting on a sequence *from* creation *to* Christ helps to address the concern that a focus on the cross sacralizes violence. The cross entails suffering inasmuch as it needs to bear adequate witness to the traumatized creation, but no more. The cross does not recommend or glorify suffering for its own sake. Moreover, the fact that Christ had some choice in his suffering, while most survivors of ecological trauma do not, can also be accommodated by the ordering of this model. The ecological trauma is not chosen and not desired, but Christ's suffering is chosen and is desired because Christ commits to bearing witness to the traumatic wounds of the world.

Divine Companionship

A second option for a theological response to ecological trauma is a model of divine companionship. Specifically, Ruth Page contends that the deep-seated ambiguity that characterizes the natural world can be ameliorated by considering the way that God "companions" creation.[85] "Companioning is a way to act in the midst of Ambiguity," she writes.[86] Nothing in creation is excluded from this divine companionship, and the companioning persists in bad times as well as good. Furthermore, God's companionship is entirely for the benefit of the entity concerned, rather than being driven by any external or ulterior motivations. As Page puts it, "companionship is about creation as creation"; creation is accompanied on its own terms.[87] So what difference does this companioning make? Everything and nothing, says Page: the world is "as complex as before" but the intrinsic value accorded to it by divine presence means that it has "changed completely."[88] The material reality of creation's suffering does not change, but the fact that this suffering occurs is honored and dignified by divine attention.

Page's companionship model is helpful for responding to ecological trauma because it does not try to offer solutions. Instead, it affirms a way to act in the midst of ambiguity. In much the same way, witnessing constitutes a way to act in the midst of trauma. Page's concern for "creation as creation" also accords with trauma theology's commitment to take reality as it is found. Moreover,

Page partakes in the sort of soteriological reconfiguration expected of a trauma theologian: for her, salvation consists in the discovery that a companionable relationship with God has been, and will be, ongoing.[89]

Yet Page's companionship model is cast in terms of a generic divine presence. As such, it could be further strengthened by a theology of the incarnation, where God is already understood as Emmanuel, God with us.[90] Moreover, God's enfleshed presence in Christ enables a much more tangible form of divine companionship. This is precisely what the model of Christic witnessing seeks to do.

Co-Suffering

Johnson and Christopher Southgate pursue a natural extension of Page's copresence model by exploring divine co-suffering with creation. Following the initial work of Arthur Peacocke and Jay McDaniel in this area, the co-suffering model is frequently applied, not just to humanity, but to all suffering and dying in the nonhuman realm.[91] Johnson insists that "the most fundamental move theology can make . . . is to affirm the compassionate presence of God."[92] Co-suffering, she maintains, is "one of the most significant things theology can say."[93] Meanwhile, Southgate, in his endorsement of a co-suffering model, explains why co-suffering is sometimes deemed so fundamental. He writes:

> I cannot pretend that God's presence as the "heart" of the world
> takes the pain of the experience away. . . . I can only suppose that
> God's suffering presence is just that, presence, of the most profoundly
> attentive and loving sort, a solidarity that at some deep level takes away
> the aloneness of the suffering creature's experience.[94]

The attraction of the co-suffering model can be hard to articulate, especially since it sounds so futile. It does not take the pain away. Yet God's co-suffering ensures that creatures are not abandoned to die alone; they are connected to God and therefore accompanied in their anguish.[95] If co-suffering is understood Christologically, this divine solidarity indicates that there is someone— Christ—who understands ecological suffering from the inside.

But Niels Gregersen and Edwards go one step further: for them, Christ's co-suffering with creation is also redemptive. As Gregersen writes, "the death of Christ becomes an icon of God's redemptive co-suffering with all sentient life."[96] God's participation in creaturely suffering also achieves redemption by a sharing of God's life-giving power. "It is not enough," writes Edwards, "to say that God is lovingly present with suffering sentient creatures. Both God suffering with creatures and the resurrection promise to them are essential."[97] Gregersen and Edwards therefore transform the emphasis on co-suffering with a promise of resurrection for all creation: the potentially tragic vision of co-suffering is invigorated by an eschatological redemption. To be clear, neither Johnson nor Southgate stop at co-suffering either. Johnson proceeds to an investigation of so-called "deep resurrection," and Southgate includes co-suffering as just one of four components of a compound evolutionary theodicy.[98] What is interestingly different, though, is that Gregersen and Edwards imply that co-suffering is inherently redemptive, whereas Johnson and Southgate suggest that co-suffering ought to be supplemented by redemption.

However, in the case of ecological trauma, there are problems with both versions of the co-suffering model. Co-suffering alone seems futile and helpless; but redemptive co-suffering seems, not only to rush too quickly to a solution, but also to make further suffering a necessity for salvation. This tension between remaining with present suffering and indicating future resolution is what lies at the heart of all trauma theologies: neglecting the former can betray the experiences of survivors, yet ignoring the latter seems to be abandoning Christian hope. Does Christ not ultimately fail in his mission if he is unable to bring about salvation and redemption?[99] An important question, therefore, is whether it is possible to conceive of a Christology that reaches beyond the passivity, and even hopelessness, of unmitigated co-suffering without risking the hollow triumphalism of a cheap redemption.

This is where the concept of cruciform Christic witnessing offers a valuable alternative to models of co-suffering. It is particularly important to recognize that the practice of bearing witness is neither static nor hopeless, as Laub's account of witnessing brings to the fore. He writes, "the listener . . . is a party to the creation of knowledge *de novo*"; they are looking for "a record that has yet to be made."[100] The survivor and the listener are not in the business of conveying a truth that is already known and mastered, like in the proclamation model of witnessing, but rather work together to unearth the trauma as they both seek to bear witness to it. In the process, the listener also comes to experience the trauma for themselves. Laub explains:

> The listener . . . by definition partakes of the struggle of the victim with the memories and residues of his or her traumatic past. The listener has to feel the victim's victories, defeats and silences, know them from within, so that they can assume the form of testimony.[101]

As such, bearing witness is something other than simple repetition. It involves creation *de novo*.[102] The result of this creation is not immediate and straightforward salvation for the survivor, but neither is it stagnation in perpetual suffering. The act of witnessing is constructive and purposeful while also remaining mindful of the dangers of any rushed redemption.

To return to a Christological key, it is this creative aspect of witnessing that distinguishes it from the relative stasis of co-suffering or copresence.[103] Christ certainly relives the sufferings of creation, but there is also something new in the act of cruciform Christic witnessing. Christ's witness is active as well as passive, creating a record that would not otherwise exist. In this way, Christ bears a kind of double witness, simultaneously recognizing both hope and loss: hope in the possibility of transformation as this new record is created, and loss as we mourn past sufferings that cannot be undone. Christ's cruciform witness stays with reality in the same way that Haraway stays with the trouble: by refusing to succumb to either naive optimism or fatalist despair.

Christ as Microcosm

A further model for understanding Christ's response to ecological trauma is the patristic and medieval notion of Christ as a microcosm, or miniature replica, of the whole cosmos. It is a theme that features in authors such as Gregory of Nyssa, Maximus the Confessor, Gregory the Great, and Bonaventure.[104] The idea builds on an understanding of human beings as microcosms. If, for example, a human being possesses existence, life, feeling, and understanding, then they can be taken to encapsulate, on a small scale, all existing, living, feeling, and understanding things.[105] However, humans ultimately fail at this unifying task, so it is left to Christ in his human nature to perform the role of perfect microcosm.[106] The cross can then be understood on the same terms. As Gregersen writes, "the cross of Christ is like a microcosm in which the sufferings in the macrocosm is both represented and lived out."[107] In this view, the cross can be understood as a small-scale mirror of the ecological trauma that occurs across the cosmos. This would accord with the idea of cruciform Christic witnessing.

Yet the notion of Christ as microcosm also reveals an important caveat about the nature of representation. There are numerous processes within the cosmos, such as photosynthesis, that cannot be directly represented within a Christic microcosm, or a human microcosm, because they do not occur within human flesh.[108] As Gregersen puts it, "it seems counterintuitive to speak of Jesus as incorporating the flying of eagles or the swimming of dolphins."[109] Hence, one of the shortcomings of the idea of Christ as microcosm is that there are many forms of suffering that are not directly borne by Christ's human body, at least not without extending our understanding of the body of Christ.[110] Nevertheless, to say that Christ *represents* ecological suffering is not to say that Christ *repeats* ecological suffering. To be a representative does not mean that you must encapsulate all the features of that which you are representing. In a model of cruciform Christic witnessing, this point about the nature of representation needs to be acknowledged. There is certainly no one-to-one correlation between the traumas of the Earth and the traumas of Christ. But there may be a minimum degree of commonality required for Christ to act as an effective witness.

God as Traumatized

One final theological approach to ecological trauma that deserves mention is the possibility that God is so completely present to the Earth and its sufferings that it is God who becomes traumatized. The trauma of the Earth is the trauma of God. As I mentioned in Chapter 2, this is the line of thinking pursued by Danielle Tumminio Hansen, Mark Wallace, and Matthew Eaton. As Wallace puts it, "God . . . is so inherently related to the universe that the specter of ecocide raises the risk of deicide."[111] In this case, God identifies so completely with ecological suffering that God is at risk of destruction.

The problem with these proposals is not so much that God suffers so extensively—there are plenty of defenses of divine passibility—but that they do not typically explain how this divine suffering relates to the historical crucifixion of Christ. Theological resources for talking about a divine response to ecological trauma already exist within Christology, without needing to identify God with all of creation. A model of cruciform Christic witnessing moves in a different direction to Tumminio Hansen's traumatized body of God or Wallace's wounded Spirit by focusing attention on Christ.

* * *

The various theological proposals engaged above help to refine my model of cruciform Christic witnessing. This model shares in the idea the Christ is present to, suffers with, and shows solidarity toward the survivors of ecological trauma, but Christic witnessing also has some distinctive features. Contra Rolston, Christ's suffering follows from, rather than preempts, ecological suffering. As opposed to Page's generic divine companionship, God's witness is envisaged in a specifically Christological form. In contrast to the passivity implied by notions of co-suffering, Christic witnessing involves taking an active stance. Unlike the traditional idea of Christ as microcosm, Christ does not already encapsulate all ecological processes. And Christic witnessing does not follow those who completely identify the divine with the Earth. Instead,

Christ's witness to ecological trauma is enfleshed, responsive, proactive, representative, and involved.

Conclusion

In this chapter I have argued that, in the wake of ecological trauma, it is Christ who bears witness to the wounds of the world. This Christic witness contrasts with models of witnessing that rely primarily on proclamation or imitation. Instead, the key feature of this witnessing is an insistence on staying with the trouble, remaining with the suffering. At the same time, the very act of witnessing brings about something new: a communication that is not otherwise possible. In this way, the Christic witness inhabits a tension between a need to honor the present, and a need to offer hope for the future.

A question that has been lingering in the background throughout this chapter is: what does such witnessing achieve? In some ways, this might be the wrong question to ask; it sounds too functional and too impatient to move on. Yet, at the same time, a consideration of this question reinforces the point that the practice of bearing witness is not claiming to bring about redemption or salvation in any straightforward way. Rather, in the words of Haraway in the epigraph to this chapter, bearing witness is about learning to be truly present, entwined in myriad unfinished configurations. This is how we can begin to process the traumatic rupture of communication.

In the next two chapters I will be extending and expanding this model of Christic witnessing in response to the other two traumatic ruptures—to flesh and to time. In Chapter 5, I will argue that Christ's incarnation and incorporation of flesh is central to his ability to bear adequate witness to ecological trauma; in Chapter 6, I will consider the necessary duration of this Christic witness in response to the traumatic rupture of time in the ecological context.

5

The Rupture of Flesh: Deep Incarnation and Enfleshed Witnessing

Our bodies are the texts that carry the memories and therefore remembering is no less than reincarnation.[1]
— KATIE CANNON, "Womanist Perspectival Discourse and Cannon Formation"

On 11 March 2011, the most powerful earthquake ever recorded in Japan triggered a major tsunami. Some 20,000 people were killed. Entire settlements were obliterated. The disaster also triggered an explosion at the Fukushima nuclear power plant, spreading dangerous levels of radiation across the surrounding landscape. The result was the worst nuclear accident since Chernobyl. Fukushima became a zone of unfolding and ongoing traumas for humans and nonhumans alike.

In 2016, two artists, Eva and Franco Mattes, traveled inside the radiation exclusion zone to take photographs. They captured images of a wide range of surfaces, from floorboards and linoleum prints to pavements and grassy lawns.[2] All of the images are freely available online, with the intention that they are used by artists, designers, and architects as backgrounds and textures in other works. The project, entitled *Fukushima Texture Pack*, enables material surfaces from a site of continuing trauma to be incorporated into new creations in other places. These new creations then bear silent witness to the traumas of Fukushima. As Franco Mattes puts it, "these textures do not depict the disaster; they come from the disaster, evoking it, and carrying a sense of incomprehension of the event."[3] In other words, the texture pack does not directly proclaim or imitate the earthquake, the tsunami, or the nuclear explosion, but offers instead a set of material witnesses to these traumas. By choosing to use these textures, people will be evoking what occurred in a tangible form.

My aim in this chapter is to address the second of the three ruptures that characterize ecological trauma—the rupture of flesh. The *Fukushima Texture Pack* offers one example of how this might be done: by encouraging the incorporation of a material, or fleshly, witness to traumatic suffering. In many respects, the rupture of flesh is the most obvious symptom of trauma. The wounding of human flesh, for example through warfare or abuse, often lies at the heart of human trauma. The subsequent trauma process, during which survivors struggle to articulate their pain, or experience flashbacks of the original wounding event, is what gives rise to the further ruptures of communication and time. In the same way, ruptures of ecological flesh—lacerations to creatures, landscapes, or the Earth itself—are central symptoms of ecological trauma.

In Chapter 4, I introduced the idea that Christ bears witness to traumatic ruptures in the ecological realm. Here, I argue that it is Christ's incarnation in the *flesh* that enables Christ to serve as an *enfleshed* witness to these *fleshly* ruptures. The trauma studies literature insists that the body, or the flesh, is the locus of trauma. Likewise, ecological flesh is the locus of ecological trauma. Meanwhile, deep incarnation theologians interpret *sarx* (flesh) in the Johannine prologue as an indication that Christ's incarnation is about his assumption of materiality. Christ must be more than just a distant representative of the wounded world if survivors are to feel a sense of divine solidarity, but Christ cannot be pancarnate in all that exists without the risk of encouraging acts of violence and trauma from the inside. Instead, Christ is in the process of incorporating fleshly ruptures that bear witness to the wounds of the world. Like the *Fukushima Texture Pack*, Christ's flesh is a material witness from a zone of ongoing trauma.

Bodies, Flesh, and Trauma

Medical work on human trauma indicates that the body is central to traumatic experience. Trauma survivors typically experience a rupturing of their bodily integrity. Indeed, the Greek word "trauma" was originally used to denote a physical wound, a bruising or breaking of a material body.[4] Yet the body is involved, not just in the initial wounding event, but also in the subsequent trauma process. What trauma psychiatrists document in cases of human trauma is that there are important connections between mind and body such that trauma is recorded somatically.[5] Traumatic events are often too extreme to be "processed" in the moment, meaning that they do not get properly archived in the brain's memory system. Instead, ongoing psychological and physiological responses—such as a recalibration of the brain's "alarm system," an increase in stress hormone secretion, and alteration to information processing filters—become "trapped" in the body.[6] A flashback can also reactivate these brain-body connections, increasing heart rate, raising blood pressure, and inducing sweating.[7] The memory of trauma is inscribed on the body. As Bessel van der Kolk makes clear in the title of his book, "the body keeps the score."[8] Bodies are therefore the locus for trauma, not just in terms of the initial physical wounds, but also as a material record of, or witness to, the whole trauma process. The epigraph to this chapter, from the theologian Katie Cannon, reiterates the point: "our bodies are the texts that carry the memories," she says.[9] In fact, bodies are so central to the experience of trauma that we might even say that part of what it means to be embodied is to be vulnerable to trauma.

In parallel fashion, ecological bodies are also the locus for ecological trauma. Trauma psychiatrists typically work within a medical framework that focuses on human trauma, and it would be a mistake to assume that all human symptoms of trauma can be translated to other parts of the natural world in a straightforward way. Yet, as I argued in Chapter 1, the very notion of ecological trauma consists in a conscious anthropomorphism of ecological and planetary processes. It is common to speak of a body of water or the body of a mountain. Earth scientists also talk about the body of the Earth, as opposed to its surface, or the existence of various planetary bodies. These metaphors already exist, blurring the boundary between what is understood to have a literal body and what is not. The important point is that, within this anthropomorphic approach, it is perfectly possible to envisage certain ecological processes as traumas on the body of the Earth. As Louis Heyse-Moore puts it, droughts, rising sea levels, forest fires, and iceberg calving can all be considered "body memories" within the Earth system.[10] To rephrase Cannon's quotation: the bodies of the *Earth* "are the texts that carry the memories" of the traumas inflicted upon them.

However, despite the importance of bodies, this second traumatic rupture is arguably best characterized as a rupture of flesh. The relative merit of the terms body and flesh is the subject of much disagreement among theologians who write on both trauma and incarnation. For some, a shift from bodies to "mere flesh" is indicative of a reduction and depoliticization of the body, opening it up to further control and manipulation.[11] This is a valid concern, but it underestimates the body's many influxes and effluxes, and overestimates the security that can be conferred by insisting on the terminology of the body.[12] For others, bodies are always inanimate, whereas "flesh is living body."[13] This perspective reveals a preference for flesh, but also specifically excludes much of the nonhuman realm from the category of living flesh. Still others note how the word body seems to denote a complete and stable entity, while flesh allows for process and change.[14] This latter point is put especially powerfully by disability theologian Sharon Betcher: "Whereas *body* can invite the hallucinatory delusion of wholeness, and thus the temptation to believe in agential mastery and control, *flesh* . . . admits our exposure, our vulnerability one to another."[15] Betcher is expressly concerned that body is an aspirational, or even transcendental, term that projects a certain vision of corporeal completion, but consequently marginalizes disabled flesh.[16] By contrast, "flesh names a locus of flux" and interchange, making it easier to countenance "pain, difficulty, disease, transience, aging, error, and corporeal limit."[17] Flesh is a preferable term to body because it is ultimately more inclusive, allowing for entities that are incomplete or distributed, and that often do not conform to the boundaried edges or somatic perfection that is sometimes expected of bodies. Theologically, the decision to refer to flesh rather than body is in keeping with the Johannine prologue, where the *Logos* (Word) becomes *sarx* (flesh), rather than either *soma* (body) or *anthropos* (human).

Likewise, it makes sense to talk about not only ecological bodies, but also ecological flesh, as the locus of ecological trauma. Ecological flesh refers to any collection of matter, animate or inanimate, that can be imagined as undergoing traumatic wounding. This includes, for example: gunshot wounds in an animal's side as a species is hunted to extinction; the bald stumps and exposed soil of large-scale logging and deforestation; the scars left on the landscape by unsustainable strip and opencast mining; or even the growing cracks in the West Antarctic Ice Sheet, which reveal the otherwise intangible trauma of rising temperatures.[18] The wholesale inclusion of all matter under the category of flesh is, as the next section demonstrates, another resonance with certain interpretations of the Johannine prologue.[19] Ecological trauma can therefore be considered in terms of a rupture to ecological flesh.

Deep Incarnation in the Flesh

Theologies of the incarnation are well suited to a theology of ecological trauma because of a mutual interest in materiality. Deep incarnation theologies are especially helpful for a Christological response to ecological trauma because of their expansive understanding of the flesh. Several of the deep incarnation theologians already featured in passing in Chapter 4, but it is worth setting out some of their central claims here.

According to deep incarnation theology, Christ was not just an isolated human individual, but rather took on a much deeper connection with material existence in the act of incarnation. For Niels Gregersen, this connection with materiality can be traced back to scripture. The "primary testimony of deep incarnation," he says, is found in the prologue to John's gospel.[20] Here, we are told that "the Word (*Logos*) became flesh (*sarx*) and lived among us."[21] What is notable, and perhaps surprising, is that this foundational text of incarnational theology does not explicitly state that God became human. Indeed, the New Testament never actually says that God became human at all.[22] Instead, the focus is on flesh. The incarnation is a coming-into-flesh.[23] For this reason, Gregersen suggests that Anselm's famous question — *Cur Deus homo?* (Why did God become human?) — should be reformulated as *Cur Deus caro?* (Why did God become flesh?).[24]

Although John's presentation of the incarnation as becoming flesh is unequivocal, the interpretation of *sarx* is more ambiguous. Gregersen describes how John uses *sarx* in three different senses throughout his gospel: it can refer to the literal flesh of Christ's body, the generally sinful nature of humanity, or the whole realm of materiality including its frailty and its vulnerability.[25] For Gregersen, it is the third of these possibilities that is the most startling and the most profound since it suggests divine presence within basic material stuff. As he writes, "*sarx* covers the whole realm of the material world from quarks to atoms and molecules."[26] The proposal of deep incarnation, therefore, is that the Johannine coming-into-flesh is a coming-into-matter. However, Gregersen is not asserting that John himself necessarily intends to convey this third meaning of *sarx* when he uses the word in the prologue. Rather, the decision to adopt this third reading is a hermeneutical choice on Gregersen's part in light of contemporary concerns about evolution and ecology.[27] Nevertheless, this particular reading of *sarx* in the Johannine prologue is in keeping with the proposal to allow all matter to be included under the heading of ecological flesh.

This expansive interpretation of flesh also has repercussions for how we view Christ's relationship to traumatic ecological suffering. According to Gregersen, "the divine Logos . . . has assumed not merely humanity, but the whole malleable matrix of materiality."[28] Similarly, Elizabeth Johnson describes Christ's flesh as "a complex unit of minerals and fluids, an item in the carbon, oxygen, and nitrogen cycles."[29] And Denis Edwards explains how "the flesh of Jesus is made from atoms born in the processes of nucleosynthesis in stars."[30] At one level, these claims are uncontroversial. It is indisputably the case that the incarnation involved Christ's assumption of certain atoms and molecules; Christ was made of stardust just as much as the rest of us. Furthermore, given the rapid recycling of chemical elements within the human body, it is likely to be the case that atoms that were once in Christ's physical body are now distributed across the Earth. But the point here is also more substantial than such literalistic statements admit. In a regularly repeated definition of deep incarnation, Gregersen writes that Christ "conjoins the material conditions of God's world of creation at large ('all flesh'), shares the fate of all biological life forms ('grass and lilies'), and experiences the pains of all sentient creatures ('sparrows and foxes')."[31] For Gregersen, the materiality of deep incarnation involves "all flesh" in a way that allows Christ to "conjoin," "share," and "experience" all aspects of creation. It is not just that Christ's incarnation in the flesh means that Christ possesses the same atoms and molecules as other parts of the material creation, but that Christ is somehow involved with the trials and tribulations of these other material forms. Notably, this view of incarnation includes incarnate involvement in the most traumatized and seemingly most hopeless corners of creation—what Gregersen calls the *tenebrae creationis* (the "darkness" of creation).[32] Gregersen seeks "the ambiguous traces of God in the world at large" and "does not confine God to the sunny side of experience."[33] In taking on flesh, Christ simultaneously takes on material frailty, vulnerability, and susceptibility to death and decay. These seemingly hopeless corners of creation are part and parcel of the flesh of the Earth and so they are shared by Christ in the incarnation.

An Enfleshed Witness

The proposals of the deep incarnation theologians are conducive to a theology of ecological trauma because Christ's conjunction with "all flesh" allows us to expand the model of Christic witnessing from Chapter 4. Previously, I argued that Christ is a more suitable witness to ecological trauma than the Holy Spirit because it is Christ, not the Spirit, who takes on flesh, and ecological trauma is an inherently fleshly phenomenon. Here, though, it is worth explaining in a little more detail why Christ's flesh is so important for his witness to ecological trauma. In essence, there are two reasons: first, Christ's flesh enables him to share in the conditions experienced by the survivors of ecological trauma; and second, Christ's flesh continues to provide a witness even when other forms of communication have broken down.

Sharing in the Flesh

Johnson proposes that deep incarnation theology allows for an expansion of Christ's solidarity to the whole of the nonhuman creation.[34] This is what enables a model of Christic witnessing to accommodate an ecological dimension. But Johnson goes on: "Incarnation bespeaks a different form of divine presence marked by an unimaginable intensity of intimacy. It is presence in the flesh."[35] Christ can express solidarity with, and bear witness to, the non-human creation precisely because of his presence in the flesh. It is because Christ became flesh, and shared in the very substrate that is ruptured by trauma, that he can perform this function. This solidarity, or witness, in the flesh reaches a dramatic climax in the events of the crucifixion. In Johnson's words, Christ "enters the fray," "freely participating in the groaning of the flesh."[36] Moreover, this solidarity applies to the whole of creation because, according to Gregersen's preferred interpretation of *sarx*, Christ's assumption of flesh is an assumption of "the whole malleable matrix of materiality."[37] As Gregersen says, Christ united himself in the incarnation with "very basic physical stuff."[38] Christ shares in the materiality of the created world, and so Christ can show solidarity toward, and bear witness to, all traumas that occur in that material world. The wider one's understanding of flesh, the wider one can envisage Christ's witness to extend.

The crucial feature of this sharing in the flesh is that it allows Christ an internal relation to the travails of the world. It could readily be argued that God can show solidarity toward, and bear witness to, all worldly traumas without any need for the incarnation, but such a witness would only ever be an external relation. Since Christ has experienced the traumatic rupture of flesh from the inside, Christ's witness carries a greater sense of empathy and connection. As Johnson comments, this is "divine participation in pain and death from *within* the world of the flesh. Now the incarnate God knows through personal experience, so to speak."[39] Gregersen also notes how Christ can "understand human and creaturely conditions from an internal firsthand perspective, and not only from a lofty third-person perspective."[40] Witnessing is enhanced if the witness has lived through something akin to the trauma in question. Christ's internal relationship to ecological flesh, to the materiality of the world, deepens his ability to act as a compassionate witness in cases of ecological trauma.

Witnessing in the Flesh

The second important aspect of Christ's enfleshed witness to ecological trauma is that it operates even when other forms of communication have broken down. Trauma entails a rupture of communication and a crisis of representation. Anyone seeking to bear witness to trauma inhabits a precarious role since their task is to represent the seemingly unrepresentable. As

Shelly Rambo argues, straightforward proclamation or imitation are unlikely to be sufficient to bear satisfactory witness to trauma, ecological or otherwise.[41] However, an enfleshed witness still initiates communication even when cognition, language, and other forms of action have all failed. To return to Serene Jones's Christology of mirroring from Chapter 4, Christ's witness is only possible because of its enfleshed character. The survivor "cannot speak," and Christ also undergoes a "loss of speech," but witnessing is still possible because Christ is an "embodied testimony": he enfleshes the "fractured, tortured shape of . . . traumatic existence."[42] If Christ's starting point for bearing witness to a wounded world is a sharing in the flesh of the world, then Christ's flesh bears witness, not by proclaiming or imitating, but simply by including the traumas of the world within his flesh. As Rambo puts it, "he bears witness in his body to the flesh of the world, marked in multiple configurations."[43]

The key reason why flesh can still bear witness in the absence of other modes of communication is because of its haptic character. For example, Rambo describes one of Christ's resurrection appearances as follows: "When he meets the disciples in the Upper Room and displays his wounds to them, he offers an invitation to glimpse and touch the flesh of the world."[44] Rambo not only emphasizes the witness that Christ bears to the wounded flesh of the world, she also indicates how we are to engage with this fleshly witness: through touch. Ruptured flesh is encountered haptically. Unlike sight, which tends to be the dominant metaphor for knowledge and has a habit of objectifying what is seen as if it is fully understood and mastered, touch allows for apprehension without control.[45] Touch could even be described as the most fundamental of the senses because it is what exposes the body to the world.[46] Christ's witness to trauma cannot necessarily be immediately seen, or known, or understood, but it can be touched. This is exhibited particularly powerfully in the resurrection appearance to Doubting Thomas. In this tactile encounter, Thomas reaches out a hand to touch Christ's wounds, which in turn bear witness to the wounds of the world. Even when other forms of communication have been traumatically ruptured, witness can still be borne via a haptic encounter with the flesh.

This account of Christ's enfleshed witness also bears on an ancillary question about Christic witnessing. Given that Christ himself makes scant mention of ecological concerns during the recorded events of the crucifixion, did Christ need to be conscious of what he was witnessing? Reflecting on Christ's psychology is a highly speculative exercise, but the relative success of Christ's enfleshed witness—even in the absence of cognition, language, and other forms of action—means that Christ does not necessarily need to have been conscious of the object of his witnessing. In *Witnessing: Beyond Recognition*,

Kelly Oliver makes this point clear. As she writes, "the performance of testimony says more than the witness knows."[47] This is because trauma places a limit on what can be known, understood, and spoken about. A complete account of the traumatic event is an impossibility. In Oliver's words, "the saying enacts the impossibility of really ever having said what happened"; we "bear witness to the impossibility of witnessing."[48] But the "performance of testimony" goes further than mere report because this performance is enfleshed; somatic and haptic communication can reach beyond what the witness themself consciously understands. Again, in Oliver's words, "this is not the finite task of comprehending . . . this is the infinite task of encountering."[49] As with Thomas in the Upper Room, Christ's enfleshed witness enables an encounter that goes further than what can be verbally, or consciously, communicated. Echoing and updating van der Kolk's maxim that "the body keeps the score," we might say that "Christ's flesh keeps the score."[50]

The Extent of Christ's Involvement

Thus far, this chapter has sought to show how the suggestion of Christ's deep incarnation in "all flesh" augments Christ's witness to a wounded world. But are there any limits on Christ's incarnate involvement? And can we be any more precise about Christ's relationship to creation? For example, is Christ representing, accompanying, standing for, expressing solidarity with, sharing, bearing, co-suffering with, being involved in, incorporating, assuming, being present in, or identifying with traumatic suffering in the ecological realm? Although it is quite simplistic, one way to visualize these different proposals is on a scale of Christ's increasing connection to creation. At one end, Christ is understood to be a distant representative of the traumas of the world; at the other end, he can be directly identified with them. Here, I offer greater clarity on what sort of involvement Christic witnessing should entail. The two extremes of this scale face several problems, but a process of Christic witnessing via material inscription and material incorporation provides a middle way.

Much of the literature that deals with Christ's relationship to creation is focused on soteriology: where and how does Christ need to be present for certain entities within creation to be saved? However, the focus here is different. Given the reticence of trauma theologians to talk directly about salvation, the central question is about where and how Christ needs to be present for him to bear adequate witness to the wounds of the world.

This question about the extent of Christ's involvement can also be usefully subdivided into two different issues. First, *what* does Christ need to share with a given survivor of trauma to be able to bear adequate witness?

And second, *how*—that is, via what mechanism—is this sharing achieved? The first question relates to what has been called the scandal of uniqueness.[51] In soteriological terms, the issue is that Christ does not appear to possess the same characteristics as the object of salvation. When Rosemary Radford Ruether asks "can a male savior save women?" the concern is that Christ does not share female flesh and so cannot save women.[52] By the same token, we are entitled to ask whether a human savior can save nonhumans.[53] Why should one particular creature, at one particular time, with one particular gender, race, class, sexuality, and species be the representative of all flesh?[54] If Christ could have taken any form of flesh, it seems like a remarkable coincidence that our anthropocentric and androcentric societies were sent not only a human being, but also a man.[55] In terms of Christic witnessing, the issue is similar: how can Christ bear witness to the trauma of entities that possess features he does not share? The same concern arose in Chapter 4 in relation to the idea of Christ as a microcosm of the macrocosm. The important point is that representatives do not have to possess all the characteristics of that which they are representing.[56] If they did, no representation would be possible without complete identity; all beings could only ever represent themselves. Likewise, witnesses do not have to possess all the characteristics of that which they are witnessing. By becoming incarnate in the flesh (where flesh is understood as all matter), Christ shares in enough to be able to bear adequate witness to the traumas occurring in ecological flesh (also understood as all matter). In this model of Christic witnessing, the point is not that Christ can repeat all cosmic processes, but that Christ has enough in common with the cosmos to function as its witness. The precise attributes of Christ's flesh are incidental to his ability to bear witness; the important point is that he shared that flesh.[57] The question then is *how* is this sharing achieved?

Representation

The first possibility for understanding how Christ shares in the flesh is via the notion of representation. The language of Christ as a representative is common in substitutionary soteriologies. Within such frameworks, Christ is a witness before God to the *sins* of the world. But the proposal here is that Christ is a witness before God (and before humanity) to the *wounds* of the world. Christ's wounded flesh represents the rupture of flesh in cases of ecological trauma. Numerous ecotheologians and deep incarnation theologians speak in these terms. For example, Celia Deane-Drummond expresses Christ's representative role as follows:

> The suffering flesh of Christ in some sense *stands for* all suffering and dying human flesh. Beyond that, it stands for the living, suffering creaturely world as such, and beyond that again it stands for the matter in the natural order of the earth and the wider universe.[58]

Deane-Drummond is clear that the Christic witness extends to the whole Earth, and indeed the wider universe, while the language of "standing for" suggests a relatively clear ontological separation between Christ and creation. Similarly, Norman Habel employs the vocabulary of representation. "God becomes, flesh, clay, earth," he says, and so "God is wholly in that piece of the earth called Jesus and that piece of earth, that is holy in God, *represents* all the earth."[59] This chiasmus forms the basis of Christ's ability to bear witness, but there remains a clear separation between the piece of earth that is made holy and the rest of the Earth. Likewise, James Nash writes about the incarnation as cosmic representation. He says, "the Representative of Humanity [Christ], therefore, is also the Representative of the biosphere, even the ecosphere, indeed, the universe."[60] He explains:

> The very nature of being human is to exist as *imago mundi*, a reflection of embodiment of the biophysical world.... Humans are representatives of the earth, interdependent parts of nature—and this totality is what God became immersed in through association with the Representative of Humanity in the Incarnation.[61]

On this basis, Christ is also *imago mundi*. Christ is not just a representative of God to us—the perfect *imago Dei*—and a representative of humanity to God, but a representative of the world, both to humanity and to God.[62]

However, the problem with expressing Christ's witness in terms of representation is that it risks sounding too distant from the ecological traumas that Christ is witnessing. The extent to which it is possible to imagine divine solidarity with the wounds of the world when the Christic witness is ontologically separated, temporally distanced, and spatially distinct from the survivors of trauma is minimal. Indeed, this need for a greater degree of involvement is clearly recognized by two comparatively recent examples of divine solidarity with the oppressed. In a Jewish context, in Elie Wiesel's famous question about the location of God in Auschwitz, the answer comes back: "hanging here from this gallows."[63] It is not that the man on the gallows is elsewhere represented by God, but that God is somehow here, on *this* gallows. Similarly, James Cone, in *The Cross and the Lynching Tree*, states that:

> God was also present at every lynching in the United States. . . . God transformed lynched black bodies into the recrucified body of Christ. *Every time a white mob lynched a black person, they lynched Jesus.* The lynching tree is the cross in America.[64]

Again, note how, for Cone, it is not that the cross represents the lynching tree, but that "the lynching tree is the cross." The connection is more ontological than representational. Cone does not want to go too far: "The cross of Jesus and the lynching tree of black victims are not literally the same—historically or theologically," he says.[65] Yet he insists that we must "identify Christ with a 'recrucified' black body hanging from a lynching tree" if we are to truly understand Christianity, slavery, and white supremacy in America.[66] Greater involvement is required.

Moreover, several ecotheologians echo this intuition that representation alone is insufficient for Christ to bear witness to ecological trauma. For instance, Habel writes that "wherever the land is crucified by our crimes, the crucified Christ is present."[67] Similarly, Mark Wallace also writes in this more ontological vein. He says that "Jesus' crucifixion wounds are now reopened as the whole Earth bears the marks of eco-catastrophe."[68] In both cases, Christ's crucifixion is more than just a representation of ecological trauma; Christ is profoundly present.

Pancarnation

At the opposite end of this scale of Christ's involvement is the notion that Christ is conceivably incarnate in everything and is therefore able to be completely present in all instances of ecological trauma. This proposal lies at the limit of what deep incarnation theologians are happy to endorse, although there are several thinkers who propose versions of this pancarnate approach. Laurel Schneider talks about a "promiscuous incarnation," that is, one that enters into "intercourse of all kinds."[69] This is not just a potentially provocative sexual metaphor, but also an indication that incarnation should be open, porous, multiple, and inclusive. Furthermore, she argues, Christ's own life was characterized by intercourse, multiplicity, generosity, and ambiguity: "the divinity which his flesh reveals," she says, "is radically open to consorting with anyone."[70] Schneider's promiscuous incarnation points the way toward a pancarnation in which God is co-wounded by all instances of trauma. Pancarnation involves "radical, compassionate, promiscuous love of the world to such an extent," she continues, "that suffering in any person, any body, is a wound in God's flesh."[71] Meanwhile, Catherine Keller talks about pancarnation and "intercarnations" as a way of escaping the "one and only" exclusivism of traditional Christology.[72] She writes, "a deity who is 'through all and in all' is pretty much by definition pan-carnate. But no doubt with such differences of reception, of intensity, of mattering, as to break up any simple God-world One."[73] Keller makes a crucial point: pancarnation does not necessarily have to endorse a monolithic divine presence; a pancarnate God can vary in "reception" and "intensity." Pancarnation is not so much about insisting on Christ's uniform presence in all flesh, as about remaining open to the possibility of divine presence in any piece of flesh. Lastly, Matthew Eaton proposes "a religious ecology wherein divinity is incarnate and self-expressive within all bodies."[74] He too "refuses to normalize Jesus' human body as the exclusive location or the definitive pattern of divine expression."[75] Instead, he claims, the "self-expression of divinity has no a priori restrictions, and might erupt within any body."[76] God's infinity and irreducibility is such that God could choose to become manifest in any piece of flesh.

As radical as these proposals sound, one of the central reasons why a pancarnate view is likely to be attractive is because of a patristic maxim about the relationship between incarnation and redemption. As Gregory of Nazianzus writes in his first letter to Cledonius, "for that which he has not assumed he has not healed, but that which is united to his Godhead is also saved."[77] This idea—that the unassumed is the unhealed—is also endorsed by Gregory of Nyssa, Tertullian, and Ambrose, and is regularly quoted by the deep incarnation theologians.[78] Gregory Nazianzen's primary intention with this statement was to combat an Apollinarian *Logos-sarx* Christology, in which the human soul is not assumed, and therefore the whole human is not saved. The point is that, if the Word did not assume human rationality and will, which were the cause of the first sin in Adam, then redemption is not possible in Christ.[79] It is also important to recognize the logic of Gregory's maxim: he knows, or wants, the rational human soul to be healed, and so it must also have been assumed by Christ.[80] But the mantra that the unassumed is the unhealed has subsequently been applied to much more than just the human soul.[81] As Karl Rahner writes:

> If anything was not assumed, neither was it redeemed. . . . But *everything* has been assumed. . . . And hence, everything, without confusion and without separation, is to enter into eternal life. . . . This is the reality of Christ, which constitutes Christianity, the incarnate life of God in our place and our time.[82]

Christ assumes *everything* and hence Christ redeems *everything*. Again, the principal concern for these authors is soteriological. But a parallel statement can equally be applied to a model of Christic witnessing: that the unassumed is the unwitnessed. In the context of ecological trauma, we can imagine that Christ assumes all flesh, including all traumatized ecological flesh, in order to bear witness to it. This would certainly address those concerns that suggest that a representational approach to witnessing does not involve sufficient ontological connection for God to show solidarity with the survivors of ecological trauma.

However, the major problem with these pancarnate understandings of the incarnation is that they risk making God incarnately present as a perpetrator of trauma. If Christ has assumed *everything*, then Christ is incarnately present in, and appears to actively encourage, acts of violence that generate ecological trauma, from the chainsaws used for deforestation to the power plants that burn fossil fuels. As Gregersen writes:

> It is one thing to say that the incarnate Christ is present *in*, *with*, and *for* all created beings . . . it is quite another thing to say that God is incarnate *as* a terrorist attack, *as* a rape, or *as* a natural disaster.[83]

This is not a matter to be taken lightly. If it looks like God condones ecological trauma from the inside, then we risk removing not only any hope of divine response, but also any human motivation to try to mitigate future traumas. Various proposals have therefore been put forward to try to insulate a deeply incarnate God from endorsing violence and trauma. Richard Bauckham highlights the fact that many different modes of divine presence are on offer in scripture and argues that incarnation should be reserved as a form of divine presence of a fundamentally different kind.[84] The incarnation can be transformative *for* all creation, but God cannot be incarnate *in* all creation. In a similar vein, Christopher Southgate draws a distinction between divine incarnation and divine immanence.[85] God is not always incarnate, but God can always be present; incarnation is soteriological, whereas immanence is not. Gregersen's own solution is to assert that incarnation must involve not just divine self-embodiment, but also divine self-revelation.[86] While God is present in the terrorist attack, the rape, and the natural disaster, they do not reveal anything about the character of God. One final strategy is to propose that the processes of assuming and healing are ongoing, and that Christ will only be incarnate throughout all creation at the eschaton. For instance, Deane-Drummond suggests that cosmic body of Christ imagery is an "eschatological goal, rather than a present reality."[87] She continues, "there are hints at the presence of Christ in all that is, but the atoning elements of the incarnation are not yet complete."[88]

For a Christic witness to ecological trauma, these proposals are helpful because they steer a middle course between symbolic representation and comprehensive pancarnation. Bauckham's restriction of incarnation does still run the risk that Christ becomes too distant to bear adequate witness. In Gregersen's words, there ought to be some internal "inner bond" between Christ and the suffering creation, and not just a series of external relations.[89] Meanwhile, it is not clear whether Southgate's preference for divine immanence, which is often associated with the Spirit, is sufficiently enfleshed for God to be able to share in and bear witness to all ecological traumas. Gregersen's own distinction between Christ's enfleshment and what is revealed by Christ's presence in the flesh is valuable, and echoes Keller's insight that divine presence can vary in reception and intensity. Christ's incarnate presence might mean different things in different contexts.[90] The idea that the full extent of Christ's incarnation is only completed at the eschaton is also conducive to a Christic witness that wants to retain an ontological connection without the risk of actively condoning trauma.

Inscription and Incorporation

Bearing in mind the advantages and disadvantages of both representation and pancarnation, a middle way on this question of Christ's involvement with creation can be parsed in terms of material inscription and material incorporation. Inscription and incorporation can be the means by which the Christic witness shares in the flesh of the world.

Ernst Conradie describes material inscription as a way to honor and record the wounding of flesh during earthly existence. His idea draws on the scriptural notion that our histories are forever recorded in the "book of life."[91] As a result, "nothing is lost; everything remains inscribed forever."[92] The metaphor of writing and recording is arguably best understood as a material and fleshly inscription. As Conradie states, "the whole history of the cosmos is *materially* inscribed, that is, it is fixated in the three dimensions of space and the added dimension(s) of time."[93] For Conradie, the implication is that "every moment in the earth's journey is not only of ecological but also of eternal significance."[94] There is a sense here that material inscription — on flesh, on a body, on the Earth — is a product of that flesh's journey through history. To be enfleshed is to be subject to marking, wounding, and material inscription. Over time, flesh is inscribed with a record of its own history. For human beings this is clearly true, even on a relatively prosaic level: as we get older, we accumulate scars from various incidents. Yet Conradie's principal concern here is the eschatological record provided by such inscription. The threat of erasure is overcome by an appeal to a material divine memory. As Gregersen puts it, "the scars of being experienced will never be forgotten"; inscribed flesh has an enduring significance.[95] But, as Conradie himself admits, such inscription has "no soteriological thrust"; it simply remains as a witness, a record, and a memory.[96] For many theologians the absence of soteriology constitutes a shortfall in this approach, but within trauma theology's more limited focus the assurance of a witness is precisely what is needed.

Christ can therefore be imagined as the fleshly substrate on which such material inscription occurs. Christ's flesh partakes in history and so includes marking and wounding. But Christ also incorporates—that is, includes within his own body, his own flesh—wounds from the traumatized flesh of the world.[97] Through this incorporation, Christ bears somatic witness to the ruptured flesh of myriad ecological traumas. To quote Cannon again, this time in full: "Our bodies are the texts that carry the memories and therefore remembering is no less than reincarnation."[98] Christ remembers the wounds of the world because Christ reincarnates the wounds of the world by bearing them as inscriptions on his flesh. The clearest examples of Christ's material incorporation of trauma are the wounds on his crucifixion flesh.[99] These wounds stem from the trauma of the crucifixion event, but they also serve as a witness to wounded flesh in general. Christ's incorporation of ruptured flesh enables the growth of solidarity and understanding between Christ, the Earth, and ourselves. Not only does Christ's fleshly witness enhance Christ's solidarity with the Earth, but it also prompts us, as observers of Christ's witness, to gain an appreciation of the trauma experienced by the Earth.

In analogous fashion to the *Fukushima Texture Pack* from the beginning of this chapter, Christ's flesh incorporates wounds that preserve a material witness to traumatic ecological suffering. Christ is not just a representative of ecological trauma, bearing witness at a distance; he is a participant in the same ruptured flesh, undergoing the same processes of wounding. But neither is Christ fully incarnate in everything. He is still in the process of incorporating the wounds of the world. Material inscription is ongoing. In this way, Christ is sufficiently involved in the traumatic rupture of flesh to serve as a witness, without completely collapsing Christ's presence into creation.

Conclusion

In this chapter, I have made the case for considering ecological trauma as a rupture of ecological flesh. The literature in trauma studies makes clear that trauma is an embodied phenomenon, yet flesh is ultimately a more dynamic and inclusive term than body. The deep incarnation theologians, who specifically stress Christ's incarnation in the flesh (where flesh is understood to include all matter), are well placed to address the rupture of ecological flesh. I therefore sought to expand on the model of Christic witnessing from Chapter 4 to emphasize the enfleshed character of Christ's witness. In joining the very substrate of creation, Christ is enabled to become a witness to the traumas undergone by all flesh. It is by incorporating traumatic wounds into his own flesh that we can say that "Christ's flesh keeps the score" of ecological trauma.

In Chapter 6, I will be turning my attention to the temporal dimensions of this fleshly witness. How long does Christ's witness last? And how can his witness respond to the traumatic rupture of time?

6

The Rupture of Time: Witnessing Anthropocene Scars

The result, therefore, of our present enquiry is, that we find no vestige of a beginning,—no prospect of an end.[1]
— JAMES HUTTON, *Theory of the Earth*

Its scarred body can never be made to disappear.[2]
— ANIL NARINE, *Eco-Trauma Cinema*

In the spring of 1788, the Scottish geologist James Hutton set off on a boat trip along the Berwickshire coast east of Edinburgh. At Siccar Point he uncovered a series of vertical slabs of hard, grey, sedimentary rock that were truncated and overlain by horizontal beds of red sandstone. Assuming that geological processes of sedimentation, mountain building, and erosion have always occurred at a similar rate, Hutton realized that vast stretches of time were needed to produce the formations he saw in front of him. The first set of rocks must have been deposited, tilted, and eroded, before the second set could be layered on top. The whole process must have taken tens of millions of years. As Hutton commented, he discerned in these rocks "no vestige of a beginning,—no prospect of an end."[3] He had discovered "deep time."[4]

In recent years, academics, cultural commentators, environmentalists, and politicians have all taken up the concept of deep time as a way of thinking about not just the geological past but also the planetary future. Within environmental literatures, deep time has been invoked by numerous thinkers to draw attention to the long-term impacts of human activity in the present.[5] In particular, the suggestion of the Anthropocene, the so-called age of humans, marks an apparent turning point in Earth history.[6] Human influence on the planet is now so pervasive that scientists have been considering whether a new geological epoch has begun—an epoch in which humanity is permanently altering the future trajectory of the Earth system. We are potentially leaving a signal in the rock record that will last for as long as the Earth itself. Anyone in the future who looks at a biography of the Earth will be able to divide time into before and after. The current ecological crisis is leaving an imprint in deep time.

In this chapter, I turn to the last of the three ruptures that characterize ecological trauma: the rupture of time. The hypothesized advent of the Anthropocene marks just such a rupture. More specifically, the ecological rupture of time refers to the way that linear timelines of gradual change are interrupted by both the *recurrence* of various ecological phenomena and the *permanence* of ecological destruction. Building on previous chapters, my aim here is to indicate how a model of enfleshed Christic witnessing can respond to these temporal dimensions of ecological trauma. Given such nonlinear temporalities, it is helpful to pursue Christologies that eschew theological narratives of lockstep progression or straightforward healing. First, a focus on helical time can grapple with the *recurrence* of traumatic phenomena in the ecological realm. A recapitulative Christology can bear witness to the temporal returns that constitute traumatic experience. Second, a deep time Christology, which understands the incarnation in the context of planetary time, can speak to the *permanence* of traumatic ecological wounding. In particular, I argue, the persistence of the wounds on Christ's resurrection and ascension body constitutes an indelible witness to the ecological trauma of the Anthropocene.

The Traumatic Rupture of Time

The rupture of time refers to a wide variety of temporal distortions and nonlinearities that accompany traumatic events. In cases of human trauma: past memories can be fragmented by the trauma process; the present is often interrupted by flashbacks; and the future can become impossible to imagine.[7] As Cathy Caruth explains, the repetitive character of traumatic experience is because "the event is not assimilated or experienced fully at the time, but only belatedly, in its repeated possession of the one who experiences it."[8] The elapsed time between the original repression of the event and its subsequent return is called its latency.[9] Hence, human trauma is often characterized by the *recurrence* of traumatic symptoms.[10]

Alongside these repetitions, the other temporal feature of note is the *permanence* of traumatic wounding. For many trauma survivors, symptoms persist for the rest of their lives. Judith Herman refers to this as the "indelible imprint of the traumatic moment."[11] Even in cases of post-traumatic growth and healing, the impact of the trauma often remains in one form or another. Both the *recurrence* and the *permanence* of traumatic wounding disturb notions of straightforward progress or incremental healing. It is simply not the case that time heals all wounds.[12] Instead, traumatic wounds return and persist, forcing survivors to learn how to live with, rather than necessarily recover from, trauma. Linear time is ruptured.

In the ecological realm, analogous temporal distortions attend traumatic events. In terms of *recurrence*, repeated disasters can be symptomatic of wider ecological traumas. For instance, Hurricane Katrina was not just a one-off event for New Orleans.[13] The 2005 hurricane was preceded by Hurricane Betsy in 1965 and followed by Hurricane Isaac in 2012 and Hurricane Ida in 2021. Hurricanes have repeatedly struck New Orleans like a series of unnerving flashbacks and, as anthropogenic climate change worsens, the frequency and intensity of these hurricanes will continue to ratchet up.[14] The underlying trauma of global warming is being made manifest in these recurring and escalating disasters. The same is also true of many other ecological processes: heatwaves, floods, wildfires, and droughts recur in similar fashion.

Climate change is also responsible for several other temporal distortions that provide compelling illustrations of the rupture of time in the ecological realm. As Stefan Skrimshire notes, climate change invokes "vast temporal distances, lags and discontinuities."[15] Much of the climate system is characterized by nonlinear feedback loops such that traumatic symptoms only appear after long lag times, intruding once crucial tipping points have been crossed. The melting of the Greenland ice sheet, for example, could already be locked in due to historical greenhouse gas emissions, even though the calamity will take thousands of years to unfold.[16] The reality of this trauma currently lies latent. As Skrimshire explains, "many future catastrophes will be in part the result of emissions already committed, and thus irreversible."[17] Rob Nixon calls such distorted temporalities "slow violence." As he writes, "by slow violence I mean a violence that occurs gradually and out of sight, a violence of delayed destruction that is dispersed across time and space, an attritional violence that is typically not viewed as violence at all."[18] Nixon explicitly has in mind processes such as climate change, the thawing cryosphere, deforestation, radiation from nuclear waste, and ocean acidification.[19] The challenge presented by slow violence is that slow-moving disasters do not usually lend themselves to striking imagery or attention-grabbing narratives, making it hard to rouse public sentiment or stimulate political action. As Nixon puts it, "the insidious workings of slow violence derive largely from the unequal attention given to spectacular and unspectacular time."[20] We live in an age that venerates spectacle and privileges immediacy, so it can be hard to pay attention to symptoms of trauma that evolve slowly or only appear after a long delay. Lag times between the initiation of slow violence and its eventual effects allow for memories to fade and resistance to falter. Again, linear timelines of cause, effect, and recovery are ruptured.

Meanwhile, the *permanence* of ecological trauma is perhaps best exemplified by the suggestion of the Anthropocene. Although it has not received formal scientific ratification, the Anthropocene remains a helpful term for expressing the sheer extent to which humanity has altered the course of the entire Earth system.[21] For example, it is expected that human activities have already suppressed the next ice age in 50,000 years' time, and possibly even the following one 130,000 years from now.[22] It is also estimated that it will take around 400,000 years for atmospheric carbon dioxide to return to preindustrial levels, and as long as four million years for the Earth's species diversity to recover from the current mass extinction event.[23] In terms of human lifespans, these intervals are of the order of 16,000 and 160,000 generations respectively.[24] Hundreds of thousands, or even millions, of years is not strictly permanence, but these timespans far exceed the frames of reference for human civilizations let alone individual human actions. Furthermore, the marker that humanity leaves in the geological record—including radioactive signatures, the skeletons of domesticated animals, and slabs of concrete—will be as permanent as the Earth itself. The Earth will also retain the imprint of Anthropocene trauma in markers as diverse as the composition of the atmosphere, the planet's remaining biodiversity, deep sea sediments, Antarctic ice sheets, and the petrified deposits of the rock record.[25] It is in this sense that we can talk about the Earth "remembering" what Herman calls the "indelible imprint of the traumatic moment."[26]

The ethicist Clive Hamilton is at particular pains to point out how drastically the Anthropocene is altering the planet's future trajectory. "The course of the Earth system has been changed irrevocably," he says.[27] For Hamilton, the Anthropocene is so revolutionary because it is now the whole Earth system, and not just isolated landscapes or ecosystems, that is being reshaped by human activities. He describes the "suddenness, severity, duration and irreversibility" of the Anthropocene, arguing against those who understand the Anthropocene as a progression or evolution of previous human impacts on the environment.[28] He continues, "the point of proposing a new geological epoch is that we are witnessing not continuous change but rupture . . . that is *permanent*."[29] The current anthropogenic release of carbon dioxide from deep geological storage will precipitate, he maintains, "changes that have everlasting consequences."[30] This is a radical rupture in the planetary timeline.

There are also two further senses in which the rupture of time in the ecological realm might be understood. First, there is the fact that our very measurements of time are being disrupted by ecological trauma. We cannot necessarily continue to rely on ice cores to provide records of climate history because the ice sheets may well be destroyed by a warming world.[31] Similarly, radiation from nuclear weapons is interfering with radiometric dating techniques.[32] The rupture of time therefore includes the erasure of the Earth's archives.[33] Second, human resource extraction involves a literal rupture of time. As Richard Irvine makes clear, there is a "temporal disjuncture inherent in our extraction of resources."[34] We are uprooting ancient organic matter from its proper place in time and injecting it into the contemporary atmosphere. When we use fossil fuels, we extract them not just from the ground, but also from the deep past and the deep future.[35] We ignore both the hundreds of thousands of years that they took to form, and the hundreds of thousands of years for which they will have an impact on global climate. As Irvine has it, "our present is a disinterred present. It draws on resources long in formation, paying its debts forward into the future fossil record."[36] Such hubristic combustion can only ever happen once in Earth history. It creates a rupture in time.

Christology in Trauma Time

We have seen how the traumatic rupture of time in the ecological realm involves both *recurrence* and *permanence*. But how might a trauma-sensitive Christology be articulated in light of these temporal ruptures? How can a model of Christic witnessing as trauma response be expanded to account for these distortions in time?

Within her trauma theology, Shelly Rambo's main temporal concern is to unsettle purely linear understandings of time. We tend to think of past, present, and future as progressive but, in the context of trauma, she says, "the past event . . . enters into the present in a way that confuses a trajectory of past, present, and future."[37] Rambo's contention is that a lockstep Christological narrative of incarnation, *then* crucifixion, and *then* resurrection is limited in its ability to speak to those who suffer ongoing trauma.[38] All too frequently, notions of healing and salvation adopt this linear narrative of progress. But such a progression is not what we, or the world, experience when we face the disjointed reality of traumatic suffering. As Rambo writes:

> A linear reading of cross and resurrection places death and life in a continuum; death is behind and life is ahead; life emerges victoriously from death. This way of reading can, at its best, provide a sense of

hope and promise for the future. But it can also gloss over the realities of pain and loss, glorify suffering, and justify violence.[39]

The desire for a fresh start is perfectly understandable, says Rambo, but new beginnings can also be seductive.[40] Survivors are made to feel as if they must pretend that everything has been fixed when their reality may be rather different. Not only is the progression from crucifixion to resurrection often unrecognizable to sufferers of trauma, but it can also mask the suffering itself by providing false hope. The rhetoric of rebuilding, restoring, recovering, and even resurrecting can become oppressive if these processes do not resonate with the experiences of survivors.[41] At its worst, this desire to progress can result in Christian triumphalism and theological supersessionism. "As long as Christian theologians are wedded to a linear reading of salvation history," says Rambo, "they will be unable to address the disruptive temporality in trauma."[42] Theological redemption understood as inevitable improvement is the opposite of a witness that remains in the aftermath of trauma.

In her own project, Rambo turns to the aporia of Holy Saturday as a way to interrupt problematically linear Christologies. Drawing on the writings and visions of Hans Urs von Balthasar and Adrienne von Speyr, she highlights Balthasar's description of Holy Saturday as "timeless."[43] For Rambo, Holy Saturday is not simply a step along the road between Friday and Sunday. Rather, the Saturday is a "suspension of time," an opening where the recurrent and persistent symptoms of trauma can be recognized.[44] Balthasar reinforces this point when he says of Christ that "no one saw the hour of your victory."[45] The apex of the linear narrative of redemption is missing; we have no access to this central moment in time. The "timelessness" of Holy Saturday is also apparent in our everyday lives inasmuch as our existence seems to permanently occupy such an interstitial moment, caught between the now and the not yet of theological redemption.[46] But Balthasar and Speyr also provide an alternative solution to the pitfalls of linear time: by mapping the Triduum spatially. For them, Friday, Saturday, and Sunday are not so much progressions in time as displacements in space.[47] On Saturday, Christ *descends* into hell, a spatial rather than a temporal translation. Balthasar further reinforces this spatial imaginary when he writes of "maps of suffering" and "a landscape of pain."[48] Vignettes from the Triduum are not necessarily to be followed sequentially but can instead be mapped onto different traumatic circumstances as they occur in space.

For Rambo, a further consequence of her nonlinear Christology is that it complicates the relationship between life and death. In the post-traumatic condition, death continues to haunt life, and life is lived in the shadow of death; there is no simple progression from one to the other. As she writes, life and death "exist simultaneously rather than sequentially ... resurrecting is not so much about life overcoming death as it is about life resurrecting amid the ongoingness of death."[49] She continues, "resurrection is already taking place in the midst of things."[50] Instead of focusing on a progression from death to life, Rambo proposes that we speak in terms of the "afterlife" of trauma and the "afterlife" of the cross where "the marks of death remain."[51] It is this thinking after the cross, she suggests, that helps us to make sense of recurring symptoms and the reemergence of different forms of suffering, even after apparent healing.[52]

Rambo's Christological concerns provide an important touchstone for theological responses to the traumatic rupture of time in the ecological realm. Her nonlinear Christology also helps to challenge those who falsely promote linear accounts of improvement. For instance, a striking rhetorical deployment of a progressive narrative occurred in the aftermath of the 2010 Deepwater Horizon oil spill. BP executives pushed a nature-and-time-will-heal argument, distracting attention from the magnitude of their own mistakes, and playing down their responsibility for the long-term cleanup operations.[53] The suggestion that "time heals all wounds" can be a strategy employed by the powerful to maintain the status quo while the true temporalities of traumatic suffering remain hidden. Yet if Christ is to bear witness to ecological trauma, it is important that this Christic witness does not endorse notions of simplistic improvement or straightforward transformation. Christ must be decoupled from purely linear temporalities. As Rambo shows, this decoupling can be achieved by emphasizing the suspension of Holy Saturday, by translating Christological vocabulary into a spatial realm, or by problematizing any easy progression from death to life. The key point is that a uniform Christological trajectory of improvement has little to say to the *recurrence* and *permanence* of ecological trauma.

Incarnation and the Avoidance of Chronocentrism

Rambo's worries about linear time also find important backing among the deep incarnation theologians. Niels Gregersen is clear on numerous occasions that deep incarnation theology must challenge what he calls "chronocentric" interpretations of the incarnation. Chronological time is constructed in the image of a ticking clock: separate parts interact in lockstep certainty to progress toward the future. Gregersen proceeds to describe what such a chronological account looks like in a Christological context: "first the cross of Christ, *then* the resurrection of Christ, *then* the church as the body of

Christ, and only *then*, finally, the resurrection of the children of God and the liberation of all creation."[54] But this projection of uniform progress onto past, present, and future does violence to those living with alternative temporalities, not least those who live according to the rhythms of the planet or the persisting symptoms of trauma.[55]

Gregersen claims that Saint Paul should bear at least partial responsibility for the prominence of this chronocentric scheme. Paul faced a challenge in the nascent church of how to explain the persistence of suffering and oppression after the resurrection of Christ. Why had Christ apparently failed to bring about liberation and resurrection for all? Gregersen's suggestion is that Paul plotted these events as a series of steps on a linear timeline to provide reassurance that they would still be reached.[56] Paul "temporalizes" the tensions between church and cosmos because "putting things into a temporal scheme makes unbearable paradoxes bearable."[57] Gregersen also blames modernity for further reinforcing chronocentric Christologies.[58] He questions the way in which European historicist thinking of the late seventeenth and early eighteenth centuries "has given priority to history (over against nature), historical sequences (over against space and eternity), and Jesus as an individual (over against his social identity)."[59] Modernity appears to have fostered a teleological approach to salvation that treats Christological loci in a stepwise sequence.

Whether Gregersen is correct to accuse Saint Paul and European historicist thought is not ultimately the important point. What is of most interest are the strategies that Gregersen employs to avoid such chronocentrism, because the same strategies are likely to be helpful in responding to the rupture of time that accompanies ecological trauma. In this regard, the core of Gregersen's proposal is a complete reconception of the relationship between Christ and time. As he writes:

> Both incarnation and resurrection are events and processes that (if they are true) are impinging on every moment and epoch in history, and are close to every place in the vast cosmic space. Before and after. Now and then. Here as well as there.[60]

Gregersen does not spell out the precise mechanics of how incarnation or resurrection impinge on every moment in history. But, given Gregersen's emphasis on the embeddedness of Christ within social, biological, and material processes, it is possible to imagine how Christ's presence persists and recurs in timeframes that extend beyond the thirty-three-year lifespan of Jesus of Nazareth. For Gregersen, the incarnation is deep in time as well as deep in space.

Nevertheless, the move to de-historicize Christ in deep incarnation theology is a contentious one. Similar Christologies, which lean heavily on the idea of the *Logos*, have tended to be viewed as overly metaphysical and unable to adequately relate to the Jesus of history. Gregersen, too, is aware that a complete abstraction of Christology from the life of the historical Jesus risks removing us from the particularities of specific instances of traumatic suffering. As he emphasizes, deep incarnation is "opposed to more shallow proposals of a universalist Christology."[61] It is precisely because of Christ that Christians can affirm that God is temporal at all. Yet Gregersen does still want to refocus our attention away from the chronology of history for a while. It is not so much that Christ should be thought of as timeless, but that Christ should be envisaged according to alternative nonlinear temporalities. It is by avoiding chronocentrism that Christ bears witness to the *recurrence* and the *permanence* of ecological trauma.

Helical Time: Witnessing Recurrence

Alternative Christological temporalities do already exist within the tradition. They have just been minimized by the prevalence of the linear view of salvation history. In other words, the critiques from Rambo and Gregersen detailed above, while accurate, only present a partial story. The view that Christianity and Judaism only possess linear views of time, in contrast to Greek and pagan understandings of cyclical time, is something of a caricature. For instance, Flora Keshgegian points out that the ancient Israelites possessed a cyclical view of time, as seen in repeated rituals such as the Passover, agrarian repetitions, and the rhythm of the seasons.[62] In the Middle Ages, Christians lived with various cyclical temporalities in the form of liturgies, daily and annual recurrences, and patterns of feasting and fasting.[63] Cyclical time evidently has a pedigree within, as well as outside, the Christian tradition.

However, the notion of helical time is even more suitable than cyclical time for theorizing a response to ecological trauma. Rambo, in grappling with the reconfiguration of time in the aftermath of trauma, quotes Catherine Keller on this issue: "The figure of the recapitulatory spiral . . . suggests a model for time itself: each moment in process recapitulates its history and yet adds its own fresh becoming, its flow neither linear nor circular but helical."[64] Moreover, "could this helical conception of time," asks Rambo, "provide a way of addressing the repetitions of trauma?"[65] In a helical temporality, time follows a spiral path. Proceeding along the loop of time there is still a past and a future, but, unlike with linear time, the past is not reduced to the bygone. Instead, events and processes are revisited, repeated, recombined, and reinterpreted in the present.[66] Importantly, though, this repetition is nonidentical. In contrast to cyclical time, where it is possible to imagine being stuck in an endlessly identical circle, helical time allows for transformation and change without forgetting the inevitability of recurrence.[67] For Keller, helical time is a natural corollary of her process theology, where beginnings and endings are deconstructed, and divinity is part of an ongoing movement of becoming. The spiral path of time "never begins from nothing" and "does not end for long."[68] As a result, "our beginnings and our endings are not the bookends of a linear history, but the double-edge of time itself."[69] This perspective is especially conducive to the approach taken in trauma theology, which centers on the present and brackets discussion of both the original cause and the long-term healing of trauma. But the advantages of helical time are not restricted to those who share Keller's metaphysical starting point; other parts of the Christian tradition also propose similarly recursive temporalities.[70]

Within Christology, helical time is particularly prominent in the idea of recapitulation.[71] Recapitulative Christology is associated with Irenaeus of Lyons, although it has been taken up recently within deep incarnation theology by Denis Edwards. The idea is that all elements of the old creation are recapitulated, or reinhabited, by Christ in the new creation.[72] Christ is the new Adam and Mary the new Eve.[73] Adam's disobedience on the tree of knowledge is revisited and relived by Christ's obedience on the tree of the cross.[74] The nonidentical yet repetitive nature of Christ's actions fits within a helical conception of time. Irenaeus also describes how the cross of Christ recapitulates a cruciformity that is present within creation, thereby anticipating Holmes Rolston's much more recent notion of the cruciform creation.[75] Edwards elaborates:

> [Irenaeus] sees the cross as inscribed across the whole creation, reaching across the sky and into the depths of the earth. . . . The cross makes fully visible the cruciform activity of the Word of God, who acts invisibly in the length and in the breadth of all creaturely reality.[76]

For Edwards, the Word's cruciform presence is a reassuring reminder that God accompanies the whole of creation in its suffering. If creation occurs through Christ, the Word of God, then it is perfectly natural for Christ to have left a cruciform watermark in the world, which is subsequently recapitulated in the historical crucifixion of Jesus of Nazareth. Christ is present in everything, and Christ recapitulates all the suffering that occurs throughout creation. In Edwards's words, it is the cross that "makes fully visible" the wounds of the world.

The notion that Christ revisits all parts of creation, including those that are suffering ongoing trauma, enables Christ to bear witness to the recurrences of ecological trauma. Just as hurricanes, heatwaves, floods, wildfires, and droughts continue to recur, Christ inhabits these repetitions, draws them to our attention, and bears witness to suffering that returns. When ecological traumas such as natural disasters recur in the same location, Christ enters these cycles of repetition and recapitulates these events. Importantly, though, Christ does not repeat the violence of these disasters in an identical fashion, but reinhabits them nonidentically. As such, he maintains a crucial balance: offering seeds for the possibility of change without eliding ongoing traumas. In Keller's words, "redemption repeats the past as different."[77] Christ's recapitulative witness inhabits a helical temporality.

There are, however, a few concerns with Irenaeus's scheme that also need to be borne in mind. First, for Irenaeus, and for Edwards, it is not just the case that Christ revisits and recapitulates Adam's trauma; Adam is also modeled on Christ.[78] On this view, the cruciformity of creation also flows from the cruciformity of Christ. However, as I explained in Chapter 4, the problem with envisaging the ordering in this way is that the wounding of creation is made into a necessity to adequately follow Christ. If a recapitulative Christology is to bear witness to recurring ecological traumas without insisting on the necessity of such traumas, then the sequence must be restricted to Christ's recapitulation of creation. Second, Irenaeus still places recapitulation within a framework of progress and growth toward an eschatological *telos*.[79] Such an insistence seems to reinscribe the sort of linear temporality that is so problematic. Irenaeus's teleological emphasis therefore needs to be softened if his recapitulative Christology is to be applied to ecological trauma. Third, Irenaeus's language often sounds too triumphant to be fully endorsed by trauma theologians. He says, for example, that Christ will "vanquish in Adam that which had struck us in Adam."[80] But the notion that Christ's recapitulation "vanquishes" ecological trauma is at odds with the recursive, persistent, and fragmentary nature of trauma. Finally, when Irenaeus speaks of recapitulation in Christ it is not always clear whether he means the whole of creation, or just humanity. Edwards is convinced that the whole of creation is included, but, in other parts of his corpus, Irenaeus sounds worryingly anthropocentric.[81] Christ needs to recapitulate all of creation if he is to bear witness to recurrent ecological traumas. Nevertheless, these concerns do not prevent Irenaeus's recapitulative motif being employed in a more guarded fashion in the contemporary context. Christ's recapitulation of ecological suffering within a helical conception of time can still allow Christ to bear witness to the recurrences of ecological trauma.

Deep Time: Witnessing Permanence

If a helical conception of time primes Christology for bearing witness to the *recurrence* of ecological trauma, then the parallel notion of deep time holds the potential for a Christic witness to the *permanence* of ecological trauma. For deep incarnation theologians, the concept of deep time is naturally appealing. As Gregersen writes, "if incarnation is deep in flesh, it is also deep in time."[82] Moreover, if Christ is understood on a geological timescale, then Christ will also be able to bear witness to wounding that is as permanent as the Earth itself. This is a Christology that will be able to engage with the temporal trauma of the Anthropocene. The challenge is to find ways to articulate the temporal depth of Christ's incarnate presence.

One way in which deep incarnation theologians have performed this temporal extension is by emphasizing the genealogies of Christ.[83] These genealogies achieve a "domestication" of deep time through a process of "nesting": biographical time is gradually connected to deep time via familial and ancestral ties.[84] However, such lineages are a double-edged sword with respect to the near permanence of geologically deep time. On the one hand, they extend the relevant Christic temporality far beyond the lifespan of Jesus. But, on the other hand, even these genealogies pale in comparison to the age of the Earth. These lineages do not, as some literalist readers once assumed, connect Christ to the beginning of the world. Instead, they embed Christ within the evolutionary development of life on Earth, which is itself embedded in an even larger geological timeframe. The genealogies of Christ point to the depth of human history, but not necessarily the depth of planetary time.[85]

A second strand of deep incarnation thinking has achieved temporal extension by focusing on *Logos* Christologies. We learn from the prologue to John's gospel that the Word (*Logos*) is not only to be equated with Christ, but also plays a part in creation itself: "all things came into being through him."[86] The *Logos* is present throughout creation before (and presumably after) the life and death of Jesus of Nazareth. *Logos* Christologies therefore succeed in greatly expanding our temporal horizons; Christ is shown to be relevant to the whole created order before (and presumably after) the existence of the human species. Yet theologians who follow this view ultimately hold that the *Logos*, that is Christ, the second person of the Trinity, is eternal, and eternity is not the same as deep time. The geological concept of deep time draws attention to the extended timescales of the Earth, but it is not referring to eternity. Several ecological traumas are likely to last for as long as the Earth itself, but they are not ultimately eternal unless they are somehow taken up into divinity. In directing our attention straight to the eternal, *Logos* Christologies are only partially helpful for enlarging the Christological field of view. If the genealogies of Christ do not quite extend the timeframe far enough, by remaining stuck within deep human history, then *Logos* Christologies extend the timeframe in the wrong direction, by leaping straight to eternity. *Logos* Christologies largely fail to dwell on the importance of planetary time.

Yet a temporally extended Christology remains important for bearing ongoing witness to the relative permanence of ecological trauma. As I discussed in Chapter 4, Rambo maintains that bearing witness must be about remaining with cases of traumatic suffering, staying with the trouble, and refusing to leave. She writes, "the concept of witness, as forged through trauma literatures, offers a way of thinking about a relationship to, and responsibility for, the past in its *ongoingness*."[87] This sort of witness is "what endures, survives, or is left over after everything else falls away."[88] Witnessing as remaining involves potentially extensive temporal duration. Rambo proceeds to describe the sort of continual witness that is required: "It is a presence that takes the form of bearing with, of enduring, and of persisting. It is an accompanying and attending presence that always carries with it the marks of suffering and death."[89] In the case of ecological trauma, where the impact of contemporary ecological devastation will endure for millions of years, and the traumatic imprint of the Anthropocene will last as long as the Earth itself, we need a witness that remains into the deep geological future: a witness to God, to ourselves, and to the planet about the trauma that is currently being wrought. One candidate for bearing witness to the permanence of traumatic wounding in the ecological realm, as well as the indelible imprint of the Anthropocene, is the wounds on Christ's resurrection and ascension flesh.

Christ's Permanent Wounds

Christ's flesh is permanently wounded flesh. At the crucifixion, nails were hammered into his hands and his feet, and his side was pierced by a spear. Scripture also attests to the persistence of Christ's wounds on his resurrection body. In Luke's gospel, for example, Christ urges the disciples to "look at my hands and my feet."[90] Likewise, in John's Gospel, "he showed them his hands and his side."[91] The most compelling account of all is the story of Doubting Thomas, in which Thomas demands to "put my finger in the mark of the nails and my hand in his side."[92] There is also a prominent strand within Christianity that maintains the wounds on Christ's ascension body. For many patristic and medieval theologians, Christ's continuing wounds serve a variety of purposes: they affirm Christ's bodily resurrection; they identify Christ's body as the same as the one that was crucified; they emphasize that Christ continues to have a body in heaven; and they act as a reminder of the ongoing significance of Christ's historical crucifixion.[93] Given the functions performed by Christ's persistent wounds, some Christians have even sought to cultivate their adoration.[94] Hymnody by Matthew Bridges suggests that there are "rich wounds, yet visible above," and Charles Wesley's Advent hymn emphasizes the "rapture" with which we should "gaze . . . on those glorious scars."[95] There is some debate in the literature about whether the marks on Christ's resurrection and ascension flesh are best described as wounds or scars, but for trauma theologians the important point is simply that the marks of trauma are recorded and remembered.[96]

However, other theologians have denied the persistence of Christ's wounds. Martin Luther, for example, associated wounding with sin and refused to see any sign of sin in the resurrection body, while John Calvin claimed that the resurrection wounds were only a temporary phenomenon to convince the disciples of Christ's veracity.[97] For Calvin, it would have been distinctly preferable for Thomas to have believed based on word alone, rather than requiring material evidence. The wounds were a fleeting accommodation to human weakness rather than anything significant in themselves.[98] In Calvin's scheme, "the resurrected body cannot carry the marks forward," says Rambo, "because the vision of heaven that Calvin presents requires a glorious body unmarked by human limitation."[99] Both Luther and Calvin placed a certain understanding of divine perfection above any need for Christ's bodily continuity.[100]

For a theology of ecological trauma, a middle way must be sought. We cannot, with Calvin, dismiss Christ's wounds as a temporary confirmation of identity, because this would fail to acknowledge the permanence of traumatic wounding. But neither should we "gaze with rapture" on Christ's scars because this would risk glorifying violence and promoting further wounding. The wounds on the resurrection and ascension body simply indicate that Christ's enfleshed witness to trauma will persist and remain. The transformation that Christ inaugurates by his resurrection does not negate or elide the marks of the crucifixion, and so does not compromise his ongoing witness to trauma. The marks are not erased. The Christic witness endures into deep planetary time.

This insistence on the continuation of Christ's wounds on his resurrection and ascension flesh also implies that Christ's wounds are present in the eschaton. As discussed above, the primary concern here is a witness that will last into deep time rather than a pronouncement on eternal states of affairs. Nevertheless, this issue provokes debate, so it merits brief comment here. For example, Ernst Conradie is concerned about the pastoral implications of an eschatological preservation of wounding. He writes:

> The inscription of the history of human pain and suffering would not by itself elicit hope. It would be a source of fear, not joy. It would, in fact, constitute a harrowing, tormenting image of hell. It would give eternal duration to all the evil, suffering, and pain of history. It would open the door for a continuous ritual re-enactment of such evils.[101]

Conradie is right to be worried about evil and suffering with eternal duration. This is exactly the sort of risk that is run by those who seek to glorify and replicate Christ's wounds. But an eschatological witness to former suffering need not be the same as an ongoing reenactment of evil. When Matthew Eaton writes that "trauma will be lamented eschatologically," he is clear that this is not about insisting on hopelessness, but about making sure that unforgettable violence is not, in fact, forgotten.[102] Indeed, a diverse range of literature refers to the eschatological presence of Christ's wounds. Zechariah prophesies that, at the eschaton, "they [will] look on the one whom they have pierced."[103] In the book of Revelation, we hear of the Lamb "slaughtered from the foundation of the world," another indication that Christ's wounds are eternal.[104] Meanwhile, for disability theologians such as John Hull it is important that "the broken body on earth corresponds to the broken body in heaven," because this serves as a reminder about what heavenly perfection really entails.[105] Furthermore, if Christ's wounds are present eschatologically, then the traumas that he bears witness to will also be recorded in the eschaton. For example, Danielle Tumminio Hansen argues that rape should be remembered in heaven.[106] To honor the epistemic credibility of rape survivors, and to avoid inflicting injustice on the survivors of sexual trauma, this sort of memory must not be erased. The central point—for Eaton, Hull, Tumminio Hansen, and others—is that the eschatological redemption offered in Christ does not involve either unblemished flesh, or an elision of the record of trauma. Christ's wounded resurrection and ascension flesh is not, as Conradie fears, a "tormenting image of hell," but simply a witness that remains.[107]

Witnessing the Permanent Wounds of the World

I explained above how the geological rupture of the Anthropocene is being recorded by the flesh of the Earth. But the record kept by the Earth is deeply ambiguous. On the one hand, Anthropocene markers can be understood as legacies or monuments to the existence and activity of humanity. On the other hand, the same records can also be described as imprints of traumatic violence.[108] For some, it is reassuring that human beings are making their mark, but for others, it feels like humanity's planetary impact is being recorded as permanent scars.

Yet this scarring is itself a form of enfleshed witnessing. It is a message to the far future — if there is anyone there to read it — about the planetary trauma that is currently occurring. Our violence is being recorded by the flesh of the Earth for future generations to see.[109] Skrimshire elaborates:

> Our impacts upon the earth have a moral weight that cannot be "forgotten" either with the promise of mortal erasure or with the promise of reconciliation/healing . . . the Anthropocene as a "scar" . . . has the power to deepen a sense of genuine loss, grief, and wonder at the impact of human actions upon the earth, by opening up a vastly wider temporal imagination.[110]

In this sense, the Earth, as the survivor of ecological devastation, is bearing witness to its own trauma. As Anil Narine says of the planet in the epigraph to this chapter, "its scarred body can never be made to disappear."[111] Whether we think about deforested hillsides, empty glacial valleys, or opencast mines, there is no hiding from the scarred flesh of the Earth. We live within this ongoing witness to trauma. Furthermore, Skrimshire continues, this witness "refuses the sense of an erasure of memory that would be implied in the wiping away of all materially embodied past wrongs," and, as such, "the intention is that something of the lamentable element of humanity's exacerbation of ecological decay, and extinction, is insisted upon."[112] Whatever improvements humanity makes in its treatment of the Earth, we have already left a permanent marker in the rock record that cannot be removed. The witness recorded by the Earth's flesh means that the trauma of the Anthropocene will never be forgotten, regardless of any restoration that is possible.

The idea of a scarred Earth is a visceral image. What is particularly striking, though, is that the same could be said of Christ's flesh in relation to the wounds on his resurrection and ascension body. To reword the quotation from Skrimshire above: "The impact of the crucifixion on Christ has a moral weight that cannot be 'forgotten' with the promise of the resurrection. . . . Christ's scarred body has the power to deepen a sense of genuine loss, grief, and wonder by opening up a vastly wider temporal imagination." Whatever transformation Christ brings, the trauma of his crucifixion is never undone. To adapt Narine's phrase, we might say that "Christ's scarred body can never be made to disappear." Given this parallel, Christ's permanently scarred flesh can serve as a witness to the permanent scarring of the Anthropocene. Observing Christ's wounded resurrection and ascension flesh, we are reminded that the indelible imprint of traumatic violence can never be entirely forgotten. It may be appealing to think, as in Calvin's treatment of Christ's ascension body, that the scars on the planet will ultimately disappear. But this would be to capitulate to the view that redemption entails erasure of the past. Instead, Christ's flesh remains as a permanent witness to ecological trauma in deep time.

Skrimshire also emphasizes the impact that these permanently scarred witnesses—both Christ and the Earth—can have on our own behavior in the present. They have, as he says above, "the power to deepen a sense of genuine loss, grief, and wonder," as well as to extend our "temporal imagination."[113] Skrimshire is principally referring to the wounded flesh of the Earth, but the same could be said of the wounded flesh of Christ. In the case of Christ, as Rambo underscores, Thomas's confrontation with Christ's wounds demands that he assesses the extent of his own complicity in the wounding.[114] Anyone encountering Christ's permanently wounded flesh has cause to reflect on their own involvement in acts of violence, and the timescales for which such impacts will last. This vision brings greater significance to events in the present since Christ's witness can become a resource for contemporary ethical thinking. And the same is true of the scars on the Earth: once we recognize their persistence in deep time, we are called to assess our own part in the wounding.

Here, then, in a theology of Christ's resurrection and ascension wounds, we find a Christological motif that can shed light on the temporal trauma of the Anthropocene. The indelible imprint of the crucifixion on Christ's flesh mirrors, and witnesses, the indelible imprint of the Anthropocene epoch on the flesh of the Earth. The permanent scar on the Earth and the permanent scar on Christ's body both bear witness to trauma. These witnesses are just what is needed in a Christology that wrestles with traumatic symptoms that persist deep into the planetary future.

Conclusion

In this chapter, I outlined the traumatic rupture of time in the ecological realm: from the *recurrence* of various ecological processes to the *permanence* of Anthropocene scarring. Rambo and Gregersen explain, from the point of view of trauma theology and deep incarnation theology respectively, why linear understandings of Christological activity are inappropriate for responding to these temporal distortions. Instead, I proposed focusing on helical time and deep time. Each of these temporalities has Christological precedent. Helical time is enacted by Christologies of recapitulation, as Christ breaks into the recurring spiral of time. Meanwhile, the permanence of ecological wounding in deep time is echoed by Christ in the persistence of the wounds on his resurrection and ascension flesh.

Linear schemes of inevitable human progress are out of sync with the realities of ecological recurrence and persistence. Instead, the model of Christic witnessing from Chapters 4 and 5 can be expanded once again by considering its duration.

Conclusion

Trauma tends to fracture elegant answers. It is therefore challenging to offer definitive conclusions in a book about ecological trauma. Nevertheless, to return to where I began, Mark Tansey's painting *Doubting Thomas* hints at much of my argument. In this image, we see a vivid portrayal of suffering in the natural world—a suffering that I have suggested reading through the lens of trauma. No explanations or solutions are immediately to hand. Instead, we are confronted with how to respond in the moment. Thomas, the figure in the foreground, simply pauses and bears witness to the wounds he observes. But Tansey's rendering of the scene also contains a theological dimension. The mention of the gospel account of Doubting Thomas places the suffering of Christ alongside the suffering Earth; Christ's wounds offer a reflection of these planetary ruptures, and we are left wondering what to do.

In *Witnessing a Wounded World*, I have attempted to bring the category of trauma to bear on issues that arise within ecotheology, asking how the methodological approach of trauma theologians might help to fashion new modes of attending to ecological suffering. In the process, it has become apparent that there are multiple ways in which this conversation could proceed; there are many viable theologies of ecological trauma. I have pursued one potential interaction between these theological sub-disciplines to show that such a conversation is not only possible, but also fruitful. Specifically, I opted to understand ecological trauma as applying to Earth systems and processes as part of a conscious anthropomorphism. As I explained in Chapter 1, this move is a deliberate concession to our limited imaginative capacities as we seek to form better relationships with the rest of the natural world.

There are, of course, both advantages and disadvantages to viewing ecological suffering through the lens of trauma. On the one hand, it problematizes any attempt at a comprehensive systematic articulation of the relationships between God, Christ, and the world. It also rejects wider explanations and solutions in favor of a search for theological motifs that can accompany and console amid persistent suffering. On the other hand, the category of trauma offers a fresh, and importantly parallel, approach to concerns about ecological suffering in general, and the current ecological crisis in particular. In Chapter 2, I argued that the category of trauma both serves as a placeholder for ecological suffering that is unimaginable and inarticulable, and promotes responses that eschew quick explanations and fix-all solutions. The notion of ecological trauma conveys something that we do not fully understand, gesturing toward those aspects of ecological destruction that are irreversible and overwhelming. The adoption of ecological concerns within a trauma-informed theological framework is also a very natural extension of existing trauma theologies. Given that many phenomena resist a simplistic nature/culture binary, it is to be expected that traumas rebound and reverberate between multiple subjects, human and nonhuman, individual and collective. In Chapter 3, therefore, I suggested that trauma theologians benefit from considering the ecological just as much as ecotheologians benefit from considering the traumatic.

In seeking a theological response to ecological trauma, I chose to follow a primarily Christological path. Specifically, the concept of witnessing has been central to my project, as I have sought to explain how Christ serves as a witness to traumatic ecological suffering. Over the course of Chapters 4, 5, and 6, I built up a picture of Christic witnessing in three successive stages. My analysis proceeded according to the three symptomatic ruptures of trauma that I identified in the Introduction: the rupture to communication, the rupture to flesh, and the rupture to time. In Chapter 4, I proposed that Christ begins to reestablish forms of communication, both in his life and his death, by modeling a commitment to remain. At the crucifixion he bears witness to all traumatic suffering, including ecological trauma, and especially that which has gone unspoken or unknown. In Chapter 5, I illustrated how the enfleshed nature of Christ's witness allows him to share the substrate of that which he witnesses. Christ's flesh incorporates wounds that echo the traumatic wounds of the world. And in Chapter 6, I expanded this notion of Christic witnessing again, indicating how Christ's witness recurs and persists in response to the recurrence and persistence of ecological trauma. In building this model of Christic witnessing I have drawn on a range of Christological motifs, particularly from the literature on deep incarnation. I have focused not so much on

CONCLUSION   131

Christ's redemptive or transformative role as on the part he plays in accompanying and witnessing creation.

In the rest of this Conclusion, I want to reflect further on four themes that have emerged over the course of the foregoing chapters: adaptation, hope, ethics, and Christian discipleship. These themes arise out of the preceding discussions, but also push into new terrain, allowing me to probe some of the wider consequences of my argument, and anticipating where ecological trauma theologians might go next.

Adaptation and Coping

On several occasions I have suggested that ecological trauma is not something that can be fixed, but something that must be endured. Bearing witness is a vital response, but the practice of witnessing is not the same as a cure or a remedy. As one commentator explains, "whatever else our witness might hope to accomplish, it cannot undo that which it seeks to witness."[1] There is a component of ecological suffering that is to be mourned and to be lived with, but never to be solved.

In theological terms, this focus on coping over solving has required a reconfigured approach to soteriology. Many Christian theologians would contend that Christ's primary role is that of savior. But I have followed Shelly Rambo's intuition that discussions of salvation need to be carefully rethought if they are to avoid the pitfalls of suggesting to survivors that there is a straightforward progression from crucifixion to resurrection in the aftermath of trauma. "A linear reading of cross and resurrection," says Rambo, "can gloss over the realities of pain and loss, glorify suffering, and justify violence."[2] This is not to diminish the importance of salvation, but to recognize that talking about salvation too easily may not be helpful or appropriate amid persistent pain and suffering. I have therefore placed greater emphasis on Christ's role as a witness to the traumatic wounds of the world. While I have not been offering explicitly practical or therapeutic recommendations, a theology of ecological trauma points toward a theology of coping, rather than a theology of answers.

In ecological terms, this focus on coping over solving can be mapped onto the difference between climate adaptation and climate mitigation. Climate adaptation involves adjusting to life in a warmer world, and aims to reduce vulnerability to changes that are, or soon will be, inevitable. By contrast, climate mitigation seeks to prevent or minimize global warming by either reducing emissions of greenhouse gases or actively removing them from the atmosphere. Adaptation seeks to cope where mitigation seeks to solve. In climate policy discussions adaptation was initially seen as a distraction from mitigation, but this has changed in recent decades, and a global goal on adaptation was a central part of the Paris Agreement in 2015. The pivotal insight is that adaptation and mitigation need not be mutually exclusive; work on one can often support and enhance the other. However, the possibility of adaptation and mitigation working in parallel has not yet filtered through to ecotheology in quite the same way. A lot of ecotheological literature is still very focused on solutions. Hence, one of the contributions of a theology of ecological trauma is to draw attention to the adaptation side of the agenda.

Environmental movements and manifestos that emphasize the need to adapt have some overlap with the concerns of this book. For instance, Paul Kingsnorth and Dougald Hine, originators of the Dark Mountain Project, state the following:

> We live in a time of social, economic and ecological unravelling. All around us are signs that our whole way of living is already passing into history. We will face this reality honestly and learn how to live with it.[3]

This is the first of their founding principles. According to Kingsnorth and Hine, civilization as we know it is precarious and fragile, founded on a flawed myth of progress. They want to encourage a different form of storytelling that adapts to the realities of a very rapidly changing climate. Likewise, Jem Bendell is famous in environmental circles for his proposal of Deep Adaptation. According to Bendell's 2018 essay, a climatically triggered societal collapse "is now likely, inevitable or already occurring."[4] The purpose of Deep Adaptation is to allow a space for grief, and "to find meaning in new ways of being and acting."[5] This involves reconciling ourselves to what is unfolding, relinquishing that which is irretrievable, restoring attitudes from prior to our fossil-fueled civilization, and learning resilience in the face of change.[6] The Dark Mountain Project and Deep Adaptation have both caused a certain amount of controversy, and have been criticized for promoting nihilism.[7] Yet some of their insights chime with attempts to cope in the aftermath of trauma. One does not have to agree with Bendell's prognosis of imminent and wholesale societal collapse to recognize that there are aspects of ecological change that are already locked in and are therefore irreversible and unfixable.[8] Facing this honestly and learning to live with these realities are still wise strategies. Some of the suggestions from Kingsnorth, Hine, and Bendell about grief, relinquishment, and resilience therefore resonate with a theology of ecological trauma. Specifically, the Christic witness that I have been proposing emphasizes ways of living on—or, as Rambo terms it, "after-living"—in the wake of trauma.[9] The wounds on Christ's resurrection and ascension body echo the wounds we are inflicting on the Earth. Christ's witness points us toward modes of after-living, to ways of adapting and continuing, as we seek to reconcile ourselves to the warmer world we have created.

Yet the perennial concern with a focus on coping over solving is that it encourages fatalism. This is the criticism that is most often leveled at the Dark Mountain Project and the Deep Adaptation Forum.[10] In labeling ecological suffering as traumatic and suggesting that the best we can do is to bear witness to it as it occurs, we risk resigning ourselves to climate change, mass extinction, and ecological devastation. Moreover, if we normalize ecological suffering, we could jeopardize any motivation we may previously have had to change our patterns of behavior. Several empirical studies also suggest that images of disaster and apocalypse are demotivating when it comes to ecological decision-making and conversion.[11]

It is true that part of what the category of ecological trauma seeks to recognize is that there are some aspects of creation's suffering that it is too late to prevent. These aspects must be witnessed, lamented, and protested. But it does not mean that such suffering was necessary for any greater good, or that it is not possible to prevent similar suffering in the future. Trauma theologians reject facile optimism, but they are not recommending wholesale pessimism either. The aim is to minimize harm and to discern viable and meaningful ways forward. For example, demands for mitigating greenhouse gas emissions are increasingly shifting to the idea that "every degree matters."[12] Climate change is not something that human beings either solve, or fail to solve, in a binary game of roulette. Some temperature increase is unavoidable, but some is not. As Rebecca Solnit observes, "wars will break out, the planet will heat up, species will die out, but how many, how hot, and what survives depends on whether we act."[13] Or, as Blanche Verlie explains, we should be "acting for a future that is less bad than it would have been if we did not act."[14] A recognition of ecological trauma is not a complete denial of the possibility of change.

This concern about promoting fatalism is why, in earlier chapters, I sounded a note of caution about Holmes Rolston's idea of the cruciform creation. The worry is that the suffering of creation is not only deemed inevitable, but also made necessary to fulfill its cruciform blueprint.[15] As Celia Deane-Drummond comments, "the notion of cruciform nature . . . seems to subtly endorse such suffering rather than giving the moral imperative to seek its amelioration."[16] It is for exactly this reason that the Christic witness I propose must follow from, rather than preempt or encourage, the suffering of creation. In the same fashion, recommendations to focus on adaptation must follow from, rather than preempt or encourage, irreversible ecological and social change. Moreover, just as adaptation is best considered alongside mitigation, witnessing and salvation could also be two sides of the same theological coin.

Hope in the Wake of Ecological Trauma

Allied to these concerns about fatalism is another important question: in the wake of ecological trauma, how do we speak of hope? Much has been written on the theme of hope in relation to ecotheology and the ecological crisis. Space does not permit a detailed treatment of the literature here. But it is important to indicate how some of these conversations about hope might be reframed by considering the notion of ecological trauma. As one trauma theologian notes, the category of trauma poses "chronic challenges to the language and practices of hope" because it describes a form of suffering with no apparent end.[17]

One of the commonly identified problems with the language of hope is that it can easily degenerate into false hope—hope that is cheap, triumphalist, or embarrassingly unaware of its privilege. Any hope that offers unqualified certainty about how the future will pan out, that whitewashes reality and presents simplistic solutions, or that provides an excuse not to deal with the realities of injustice is false hope.[18] As Miguel De La Torre points out, thinking about hope can easily become a middle-class pursuit that continues to oppress the marginalized.[19] The poor and the disenfranchised are so embroiled in the struggle of day-to-day existence that they do not have the luxury of sitting and waiting for God's future promises to materialize.[20] In the same vein, trauma theologians are wary of naive invocations of hope. Rambo thinks we should question the assumption that "resurrection hope points to the future," while Serene Jones writes about "a hope forever deferred."[21] Hopes of future improvement can founder on the lived realities of ongoing suffering and trauma. Instead, De La Torre proposes "embracing the reality of hopelessness" as a way to remain faithful to unending injustices.[22] As he puts it, forcefully, "do not offer me your words of hope; offer me your praxis for justice."[23] Much the same advice could be given to witnesses of trauma.

However, those who are sensitive to the persistence of trauma and the extension of injustice are not all ready to jettison hope entirely. In Chapter 4, I likened the practice of witnessing to Donna Haraway's idea of "staying with the trouble."[24] As Catherine Keller comments, "staying with the troubles, which certainly implicates future troubles, may require something very like 'hope'—the embrace of possibility—if it means to stay-with for more than one moment."[25] Keller's point is that staying with the trouble implies a duration and therefore a future. There is absolutely no guarantee of improvement, but if staying is to even be possible, then the future must be acknowledged. A seed of hope lies in the "embrace of possibility" that is realized by attending to the trouble over time. Witnessing trauma may still allow for a chastened form of hope.

The essential point is that hope is not the same as optimism.[26] With hope, there is no expectation that everything will turn out all right. Both optimism and pessimism presume they know what will happen next, but hope is an embrace of the unknown.[27] Rebecca Solnit is particularly clear on this issue. Hope is not "the belief that everything was, is, or will be fine"; instead, hope is "an account of complexities and uncertainties, with openings."[28] Real hope does not short-circuit sorrow, grief, or lament. It is only because there is uncertainty about the future that hope has anywhere to reside. In the wake of trauma, survivors and witnesses are dealing with exactly this sort of uncertainty.

The recognition that hope is complex and uncertain has resulted in a wide array of potential adjectival qualifiers. Panu Pihkala suggests "tragic hope"; Pamela McCarroll uses "authentic hope"; and Solnit herself writes about "hope in the dark."[29] Emmanuel Katangole's "hope in ruins" and Joseph Winters's "melancholic hope" convey analogous sentiments.[30] But one of the most striking proposals is Jonathan Lear's "radical hope."[31]

At the end of *Spirit and Trauma*, Rambo turns to Lear's work for an articulation of what hope could conceivably look like in the face of persistent trauma. Lear recounts the testimony of Plenty Coups, the last chief of the Crow Nation, and explains how he held on to a radical hope even after the ending of his world. Lear imagines what Plenty Coups might have been thinking:

> I recognize that in an important sense we do not know what to hope for. Things are going to change in ways beyond which we can currently imagine. We certainly do know that we cannot face the future in the same way that we have been doing. . . . We must do what we can to open our imaginations up to a radically different set of future possibilities.[32]

Plenty Coups's radical hope about the future of his people is, in Lear's interpretation, tentative and humble; it is not secured, or guaranteed, or even epistemically accessible. For Rambo, radical hope is about facing impossibility with imagination. She writes, "what distinguishes radical hope from hope is the distinctive activation of imagination in the face of an unimaginable future."[33] Imagination is the key to hope. Something like Plenty Coups's words could feasibly be uttered in the face of ecological trauma too.[34] We know that the planet is changing; we do not know what we might legitimately hope for; but we can attempt to open our minds and imagine radically different futures.

Hence, a trauma-informed theological methodology can, and should, allow for hope in response to climate change, mass extinction, and ecological devastation. But such hope is a radical hope—a hope against hope—that is far removed from optimism.[35] As Pihkala writes, hope is "a determination to continue living and to find meaningfulness, even though we cannot know how well humanity can succeed in mitigating the environmental crisis."[36] In other words, hope in the wake of ecological trauma is not dependent on success. Rather, this radical hope is maintained in the witness's commitment to remain. As I have been suggesting in this book, it is Christ's insistent refusal to abandon the wounds of the world that is the basis for his witness, and therefore the basis for Christian hope, despite the persistence of ecological trauma.

Witnessing and Ethical Action

In the preceding chapters I have often stated or implied that bearing witness to ecological trauma helps to prevent future violence. I have suggested, for example, that Christ's flesh, as a witness to the flesh of the Earth, becomes a resource for contemporary moral reflection. Our encounter with Christ's wounded witness prompts us to examine the ways in which our own habits and practices have contributed to the wounding. There is a link here between bearing witness and enacting ethics.

Yet this turn to ethics is perhaps not as straightforward as I have made out. How does one talk about practical ethical recommendations when, according to trauma theory, a complete understanding of the situation is not possible? Worse still, is there not a danger that viewing ecological suffering through the lens of trauma simply results in ethical paralysis? And, if trauma theologies are truly focused on survivors, what does it matter how anyone else responds? As I set out in Chapter 2, the advantage of a theology of ecological trauma as compared to a traditional environmental ethic is that it does not claim to have understood the *causes of* or *solutions to* ecological suffering, and has relinquished the notion of human control. So, what is the connection between witnessing and ethical action?

My strategy has been to attempt to show restraint in relation to ethical prescription. I have not wanted to repeat those ecotheological approaches to the current crisis that offer confident proclamations of what must be done, as if humanity is sufficiently engaged and sufficiently rational to fully enact such proposals. Indeed, many of our current ethical frameworks seem unable to grapple with the scale, severity, and sheer inaccessibility of ecological trauma. There is therefore nothing inherent in the practice of witnessing that dictates behavior or guarantees justice.

However, this does not have to mean that all ethics is impossible. Christic witnessing can nonetheless provide some initial pointers in terms of ethical thinking. For example, in Chapter 6, I noted Stefan Skrimshire's observation that the scarred body of the Earth—or of Christ—has "the power to deepen a sense of genuine loss, grief, and wonder," and to extend our "temporal imagination."[37] These are ethical sentiments, if not ethical edicts. Moreover, Rambo notes how the resurrection appearances of Christ's wounded body "evoke response and signal responsibility."[38] In other words, the Christic witness is a call that demands a response. As Rambo writes, "the haunting [witness] is also the transmission of the imperative to love, the call for something to be done."[39] This is notably the case for Doubting Thomas, who is called to confront his own participation in histories of violence, but it is true for us as well.[40] The key point is that witnessing forces the witness to examine their own complicity and hypocrisy; it changes them.[41] When we bear witness to the scars on Christ's flesh, we are reminded of the Anthropocene scars on the Earth, and we are forced to call to mind the ways in which our lives are dependent on the products of opencast mining or the energy from fossil fuel combustion. This power of witnessing to elicit thoughts of responsibility is also indicated by Kelly Oliver. For her, witnessing brings with it both response-ability (the ability to respond) and ethical responsibility (the obligation to respond).[42] Michael Richardson notes a similar phenomenon in relation to nonhuman witnesses. Witnessing, he says, "insists upon a response" because it "exceeds itself and calls others into relation with it."[43] Witnessing is therefore a catalyst for politics: we are confronted with witnesses that demand responses, and we are formed into communities in the process.[44] Collectives are created when people gather at sites of wounding.

This tempered view of ethical direction in response to ecological trauma is not dissimilar from Joanna Zylinska's proposal in *Minimal Ethics for the Anthropocene*. Zylinska believes we must avoid promising too much too soon, but without evading our responsibility for the actions we take.[45] A theology of ecological trauma is never going to provide policy recommendations or lists of duties, but it still commits us to the idea of accountability. There remains a radical hope, to use Lear's language, that the future might pan out differently from the past, and that witnessing to existing and persisting traumas will help to reduce future violence.

There is also a sense in which witnessing itself can be viewed as an ethical act.[46] The witness's insistence on remaining and their refusal to abandon the site of suffering is an ethical response. Such witnessing is ethical not only because it foregrounds a truth *about* suffering, but because it remains true *to* the suffering.[47] It does not betray the suffering by becoming increasingly indifferent over time, or by proffering premature explanations. To take up a posture of active and attentive witnessing is to adopt an ethical stance.

The Call on Christian Followers

So, how are Christians to respond to the model of Christic witnessing that I have been proposing in these pages? What is the place of Christianity in this vision of ecological witnessing?

My suggestion is that Christians are called, not to imitate Christ's sacrifice, but to emulate Christ's witness. As Rambo lays out, imitative models of witness risk glorifying self-sacrifice and endorsing violence.[48] But Christians can follow Christ by seeking to bear witness to ecological traumas like climate change and mass extinction. The example of Christic witnessing particularly pertains to Christians, but the wider practice of witnessing is relevant to those of all faiths and none. We are charged with bearing witness to others as they bear witness to ecological devastation.[49] We should not gloss over the suffering, nor give up on the wounds of the world. Our task is to stay with the trouble.

Yet there is a grave risk in this sort of witnessing: how are we to adequately represent the traumas of the Earth? As Stacy Alaimo writes, "speaking for nature can be yet another form of silencing, as nature is blanketed in the human voice."[50] In part, this concern is addressed by continually reminding ourselves that we should not be trying to follow a proclamatory model of witnessing. We know only too well that declaring further facts about the trauma of climate change is restricted in its efficacy to connect with people or motivate new behaviors. Further speech is of limited utility. Within the trauma literature, it is widely accepted that it is enabling survivors to narrate their own story that begins to address the rupture of communication. Therefore, our role as witnesses is not to speak for nature, but to "pay attention to what the earth might have to say."[51] The process begins with listening, so that we, as Dori Laub says, can help to set down "a record that has yet to be made."[52] We also safeguard against smothering nature in human speech by focusing on embodied and enfleshed practices of witnessing, including forms of ritual, activism, and protest.

In the Christian context, it is worth briefly noting certain congruences between the Christological call to bear witness and existing liturgical and spiritual practices. Indeed, if the body of Christ serves as a witness to ecological trauma, then it is reasonable to think that displacements of the body of Christ—in the eucharist and in the church—also function as witnesses in a similar manner. As Karen O'Donnell notes, "the principles of trauma and trauma recovery were well understood by the ancient liturgists," such that the repetition of liturgy bears witness to the trauma of the Christian story.[53] She continues, "Christian liturgy holds within it an unclaimed memory and experience of trauma."[54] In much the same way, the contemporary Christian

CONCLUSION

calendar enables an annual repetition of liturgies that foreground ecological suffering. Every year, the Season of Creation, an ecumenical initiative of the World Council of Churches, calls to mind the harm that humanity is inflicting on the Earth, while the traditional Triduum recollects Christ's cruciform witness to that suffering. These cyclical returns also align with the seasons, blending with natural rhythms of growth and decay. Such liturgical repetitions highlight the recurrence of trauma, but, importantly, they are also nonidentical in character: each return is the forging of a *de novo* witness to the same traumatic events.[55]

There is also scope for liturgical innovations that sharpen our witness to planetary-scale devastation. The psalms of lament are an important resource for trauma theologians, but there is more to be done with the ecological dimensions of the psalms that would raise up the plight of the Earth. Initiatives such as the "Stations of the Forests" resource, which tracks the traditional fourteen stations of the cross, could be regularly integrated into Passiontide services.[56] And new liturgies could be developed for remembering specific ecological calamities, such as natural disasters, extinction events, and other Anthropocene wounds. A secular example of this sort of practice is the funeral that was held for the demise of the Okjökull (Ok glacier) in Iceland.[57] The daily, weekly, and annual cycles of Christian liturgical repetition offer multiple opportunities for witnessing both the recurrence and the permanence of ecological trauma.

The role of liturgy in the attempt to follow Christ's ecological witness is especially profound in celebrations of the eucharist. Here, bread and wine, elements of the natural world, are recalled as broken and outpoured, symbols of the suffering creation. Each eucharist is a double re-membering of these elements, that is, a re-assembly and a nonidentical repetition. According to O'Donnell, the eucharist is a paradigmatic "repetition of a traumatic somatic memory," yet it is also "assembled afresh" each time.[58] Traditionally, the eucharist points to the crucifixion; for O'Donnell, it indicates the annunciation-incarnation event; according to Dirk Lange's reading of the *Didache*, it reveals the disruptive radicality of Jesus's meal-sharing tradition; and, in the context of this book, I suggest that it might signal the physical brokenness of the material world.[59] The traumatic somatic memory to which the eucharist refers could be the flesh of the Earth. Indeed, ecotheologians have long admired the emphasis on the cosmic dimensions of the eucharistic liturgy within Eastern Orthodox theology.[60] If, as Elizabeth Theokritoff suggests, the eucharist reveals the "sacramental potentiality" of all creation, then it seems plausible to think that the same sacrament, in its broken state, is echoing and witnessing the traumatic ruptures that creation undergoes.[61] Likewise, in his powerful meditation "The Mass on the World," Teilhard de Chardin envisages the whole Earth as a eucharistic offering. He writes of the "immense fragmentation" of this eucharistic host—another pointer to the parallel ruptures of the traumatized Earth and its eucharistic repetition.[62] And O'Donnell herself hints at this parallel when she writes: "The embodied experience of the Eucharist helps to create a eucharistic perspective on the natural world in which sacramental materials reflect, in their ritual use, the broken practices of the world."[63] In other words, the eucharist bears witness to the wounds of the world. What is more, this is a practice that is resolutely material and unambiguously enfleshed; words alone do not capture the offering, sharing, and consuming involved. The eucharistic liturgy is one clear way in which Christians might follow Christ's ecological witness.

Finally, given that liturgical and eucharistic practices take place in ecclesial contexts, there is good reason to think that a recognition of ecological trauma could prompt a wider ecclesiological response. Since ecological trauma can be envisaged as a type of collective trauma, where different parts of creation are taken to be subjects to different extents, it would be beneficial to think about theological responses in explicitly collective terms. As O'Donnell puts it, trauma theology should be "a community building project."[64] This ecclesiology would need to be open to participant members that extend well beyond the human. But such a space could provide a forum for multiple overlapping testimonies from the front lines of ecological devastation that are then witnessed and memorialized as part of an intentional communal activity. When we speak of the church's witness, this might include, quite specifically, conscious attempts to bear witness to ecological trauma in community. This sort of communal initiative may well be possible and valuable in secular and interfaith settings too.

Yet there remain limits to what the liturgical, eucharistic, and ecclesiological witness of Christian followers can achieve. Although we may wish that liturgies and rituals were sufficient to bring about ecological healing, ecological conversion, or ecological action, such results do not necessarily follow straightforwardly from the acts of witnessing themselves. Trauma theologians are not typically very comfortable about moving too swiftly to solutions, while hopes for conversion echo some of the more problematic aspects of models of witness that emphasize proclamation or imitative martyrdom. There is no immediate glory in bearing witness to Christ. Instead, Christians, too, are simply called to stay with the trouble.

* * *

CONCLUSION

So, where are we left? As the ecological crisis continues to unfold, the category of trauma draws attention to that which is incommunicable and tragic. A theology of ecological trauma is not a replacement for existing ecotheologies or existing trauma theologies, but it does offer a novel approach to suffering in the ecological realm. By focusing on living with the ongoing reality of trauma, this new vein of theology prompts us to think in terms of coping, reflecting, and persisting. Specifically, in Christ, we find a theological witness to the ruptures of ecological trauma. Here is a record of what has been lost and the seeds of what might be saved. This is a witness that hovers between naive triumphalism and fatalistic despair, refrains from dictating an ethics, and yet offers a radical hope. The Christian calling is to follow Christ's witness, to refuse to let ecological destruction go unrecognized, and to remain in solidarity with the survivors who suffer.

Christ's permanently wounded flesh bears witness to a permanently wounded world.

Acknowledgments

This book began life as a doctoral thesis at the University of Oxford, and I am immensely grateful to my supervisors. Celia Deane-Drummond has been a constant and calming source of wisdom and support. Her openness to my project just as the Laudato Si' Research Institute (LSRI) was starting up has made a world of difference. Graham Ward was also very generous in taking me on part way through my studies. His combination of continual encouragement and theological questioning has expanded and refined my thinking in countless ways. And Michael Oliver's initial supervisions did much to shape the direction of this project for the better. It was Michael who first suggested that I might enjoy reading the work of Shelly Rambo. Moreover, my thesis examiners, Christopher Southgate and Alister McGrath, have graciously offered valuable insights and advice both during and since my viva.

I have hugely enjoyed working with Fordham University Press as my thesis has developed into this book. I am especially grateful to Richard Morrison, Lis Pearson, Kem Crimmins, and two reviewers for the press who have all helped to nurture and improve this project in various ways. Mark Tansey has also been incredibly kind in allowing me to reproduce his painting *Doubting Thomas* on the cover. As I explained in the Preface, this image encapsulates much of what I try to say in the text. And I particularly want to thank Gabrielle Farina at Gagosian for facilitating permission to use Tansey's art.

My somewhat circuitous journey into academic theology and religion began quite some time ago. John Hughes, Jonathan Arnold, Matthew Kirkpatrick, Peter Groves, and John Bowker all helped to sow initial seeds. I have also benefited enormously from several different institutions along the way, including Harris Manchester College, Lincoln College, the LSRI, Campion

Hall, the William Temple Foundation, Pembroke College, Regent's Park College, and Oxford's Faculty of Theology and Religion. Colleagues, staff, and friends in these various places have all played a part in bringing this book to fruition. For their mentorship and support, I am particularly indebted to Tom McLeish, Alan Ramsey, and Justin Jones.

The sometimes lonely task of writing has been made much more manageable by the companionship of others. For conversations, friendship, and feedback on my work I am grateful to Mark Aloysius, Susan Bridge, Dallas Callaway, Ed Chan-Stroud, Natasha Chawla, Matthew Eaton, Cristóbal Emilfork, Preston Hill, Tim Howles, Zach Kahler, James Lorenz, Finlay Malcolm, Andrew Moore, Sarah Mortimer, Karen O'Donnell, Emily Qureshi-Hurst, David Scott, Andy Shamel, Sorrel Shamel-Wood, Bethany Sollereder, Austin Stevenson, Tobias Tanton, Raffaella Taylor-Seymour, Tom Topel, Lyndon Webb, and Oliver Wright. In addition, Penny Boxall kindly provided sage writing advice and discerning editorial guidance.

Many other friends have also contributed to this project in ways both big and small. Alas, there is not enough space to name everyone here. But I am especially thankful to Edward, Julia, Asger, and Hannah for time spent together—in person and online—during those lockdown days. I am also indebted to my family, who were incredibly patient as I pursued life as a perpetual student.

Finally, Claire, your love and your belief made this all so much more doable. Thank you.

Notes

Preface

1. Matthew's gospel records the occurrence of an earthquake at the crucifixion. See Matthew 27:51.
2. Mark C. Taylor and Mark Tansey, *The Picture in Question: Mark Tansey and the Ends of Representation* (Chicago: University of Chicago Press, 1999), 55.
3. Stephen Jay Gould even suggests that what Tansey's Thomas fails to believe is the reality of earthquakes and the theory of continental drift. See Stephen Jay Gould, *Rocks of Ages: Science and Religion in the Fullness of Life* (London: Vintage, 2002), 14.
4. Shelly Rambo, *Resurrecting Wounds: Living in the Afterlife of Trauma* (Waco, TX: Baylor University Press, 2017), 10, 19.
5. Rambo, *Resurrecting Wounds*, 9.
6. Rambo, *Resurrecting Wounds*, 17–42.
7. Rambo, *Resurrecting Wounds*, 145.
8. Rambo, *Resurrecting Wounds*, 102, 151.

Introduction

1. In this book, I focus specifically on a *Christian* theology of ecological trauma. Similarly, most of the literature I draw on—in both ecotheology and trauma theology—is by Christian theologians. However, the ecological crisis is an issue that affects those of all faiths and none, and several of the relevant resources for thinking about this subject are shared across different cultures and worldviews. A theology of ecological trauma, or a secular response to ecological trauma, would therefore be possible and valuable within other communities and traditions too. My focus on Christian theology is simply a product of my own location and commitments. I use "theology" as a shorthand for "Christian theology" throughout.

2. Susan Clayton et al., *Mental Health and Our Changing Climate: Impacts, Implications, and Guidance* (Washington, D.C.: American Psychological Association and ecoAmerica, 2017); E. Ann Kaplan, "Is Climate-Related Pre-Traumatic Stress Syndrome a Real Condition?," *American Imago* 77 (2020): 81–104.

3. Gay Bradshaw, *Elephants on the Edge: What Animals Teach Us About Humanity* (New Haven, CT: Yale University Press, 2009); Stacy M. Lopresti-Goodman et al., "Psychological Distress in Chimpanzees Rescued From Laboratories," *Journal of Trauma and Dissociation* 16 (2015): 349–66; Jessica Deslauriers et al., "Current Status of Animal Models of Posttraumatic Stress Disorder: Behavioral and Biological Phenotypes, and Future Challenges in Improving Translation," *Biological Psychiatry* 83 (2018): 895–907.

4. Robert J. Brulle and Kari Marie Norgaard, "Avoiding Cultural Trauma: Climate Change and Social Inertia," *Environmental Politics* (2019): 1–23; Leslie Stein, "Global Warming: Inaction, Denial and Psyche," in *Environmental Disasters and Collective Trauma*, ed. Nancy Cater and Stephen Foster (New Orleans: Spring Journal, 2012), 23–46. Michael Richardson also offers a definition of ecological trauma in this vein but includes nonhumans within the collective. For Richardson, ecological trauma is "a rupturing of relations that ripples through the ongoing composition of more-than-human ecologies." See Richardson, *Nonhuman Witnessing: War, Data, and Ecology After the End of the World* (Durham, NC: Duke University Press, 2024), 115.

5. Colin Murray Parkes, "Responding to Grief and Trauma in the Aftermath of Disaster," in *Death, Dying, and Bereavement: Contemporary Perspectives, Institutions, and Practices*, ed. Judith Stillion and Thomas Attig (New York: Springer, 2014), 363–78.

6. The American Psychological Association has released a series of reports on climate change and mental health, most recently: Susan Clayton et al., *Mental Health and Our Changing Climate: Impacts, Inequities, Responses* (Washington, D.C.: American Psychological Association and ecoAmerica, 2021). See also: Helen L. Berry et al., "The Case for Systems Thinking about Climate Change and Mental Health," *Nature Climate Change* 8 (2018): 282–90.

7. Panu Pihkala, "The Cost of Bearing Witness to the Environmental Crisis: Vicarious Traumatization and Dealing with Secondary Traumatic Stress Among Environmental Researchers," *Social Epistemology* 34 (2020): 86–100, 86.

8. Kaplan, "Is Climate-Related Pre-Traumatic Stress Syndrome a Real Condition?" See also: E. Ann Kaplan, *Climate Trauma: Foreseeing the Future in Dystopian Film and Fiction* (New Brunswick, NJ: Rutgers University Press, 2016). Note that there is a privilege entailed in envisaging future catastrophes as opposed to experiencing them firsthand.

9. Benjamin White, "States of Emergency: Trauma and Climate Change," *Ecopsychology* 7 (2015): 192–97, 193.

10. For some examples of theological engagements with these terms see: Panu Pihkala, "Eco-Anxiety, Tragedy, and Hope: Psychological and Spiritual Dimensions

of Climate Change," *Zygon* 53 (2018): 545–69; Lisa H. Sideris, "Grave Reminders: Grief and Vulnerability in the Anthropocene," *Religions* 11 (2020): 1–16; Hannah Malcolm, "Grieving the Earth as Prayer: A Wounded Speech That Heals," *Ecumenical Review* 72 (2020): 581–95.

11. As I discuss in Chapter 2, one of the advantages of thinking in terms of ecological trauma is that it transcends the nature/culture dichotomy. The category of ecological trauma already includes the traumas of humans and nonhumans alike. I focus here on the nonhuman and planetary dimensions of ecological trauma because they have received less attention to date.

12. IPCC, "Summary for Policymakers," in *Climate Change 2021: The Physical Science Basis. Contribution of Working Group I to the Sixth Assessment Report of the Intergovernmental Panel on Climate Change*, ed. V. Masson-Delmotte et al. (Cambridge: Cambridge University Press, 2021), 5.

13. IPCC, "Summary for Policymakers," 8.

14. IPCC, "Summary for Policymakers," 8.

15. IPCC, "Summary for Policymakers," 14.

16. IPBES, *Summary for Policymakers of the Global Assessment Report on Biodiversity and Ecosystem Services of the Intergovernmental Science-Policy Platform on Biodiversity and Ecosystem Services*, ed. S. Díaz et al. (Bonn, Germany: IPBES Secretariat, 2019), 12.

17. IPBES, *Global Assessment Report on Biodiversity and Ecosystem Services*, 12.

18. Richardson, *Nonhuman Witnessing*, 132.

19. As I discuss in detail in Chapter 3, there are very often interconnections that complicate any simplistic division between anthropogenic and natural forms of ecological suffering.

20. For example, Michael Richardson states that "like trauma more generally, ecological trauma is found not in the violence that enacts a rupturing of relations but in how that rupturing carries through into the future." See Richardson, *Nonhuman Witnessing*, 131.

21. For this reason, my definition of ecological trauma differs from the somewhat esoteric idea of geotrauma—the notion that the Earth's own evolution is necessarily traumatic. This includes, for example, the abundance of cyanide on the Archaean Earth, and the convection currents in the planet's interior, which Nick Land describes as "the anorganic metal-body trauma-howl of the earth." See Timothy Morton, *Hyperobjects: Philosophy and Ecology After the End of the World* (Minneapolis: University of Minnesota Press, 2013), 53; Nick Land, *Fanged Noumena: Collected Writings 1987–2007*, 5th Edition, ed. Robin Mackay and Ray Brassier (Falmouth: Urbanomic, 2018), 498. For Land, all human traumas are reprisals of this primordial geotrauma. There is something helpful about this sensed connection between planetary processes and human well-being, but I do not agree with Land that the Earth is constitutionally traumatized. That said, there are some authors who have employed the word "geotrauma" to describe phenomena that I include under the heading of ecological trauma, and so, in this respect, there

is some overlap between the two discourses. See Kathryn Yusoff, *A Billion Black Anthropocenes or None* (Minneapolis: University of Minnesota Press, 2018), 36.

22. I am not referring here to the way in which the word "ecological" has previously been used within the psychological literature to indicate the diverse range of interconnected social, political, and cultural factors that can have an impact on the human experience of trauma. For example: Mary R. Harvey, "An Ecological View of Psychological Trauma and Trauma Recovery," *Journal of Traumatic Stress* 9 (1996): 3–23; Mary R. Harvey, "Towards an Ecological Understanding of Resilience in Trauma Survivors," *Journal of Aggression, Maltreatment & Trauma* 14 (2007): 9–32. This is a perfectly legitimate line of enquiry, but it does not have any bearing on the natural world at large. See Pihkala, "The Cost of Bearing Witness to the Environmental Crisis," 87.

23. For example: Kaplan, *Climate Trauma*; Michael Richardson, "Climate Trauma, or the Affects of the Catastrophe to Come," *Environmental Humanities* 10 (2018): 1–19; Zhiwa Woodbury, "Climate Trauma: Toward a New Taxonomy of Trauma," *Ecopsychology* 11 (2019): 1–8.

24. Shelly Rambo, *Spirit and Trauma: A Theology of Remaining* (Louisville, KY: Westminster John Knox Press, 2010), 18; Karen O'Donnell, *Broken Bodies: The Eucharist, Mary, and the Body in Trauma Theology* (London: SCM Press, 2018), 6–7. Rambo and O'Donnell talk in terms of ruptures to words (or cognition, or language), to bodies, and to time, but I opt for the more inclusive categories of communication, flesh, and time to allow for the possibility of trauma in nonverbal, noncognitive, nonlinguistic, and nonbodily entities.

25. O'Donnell, *Broken Bodies*, 7.

26. Hilary Ison, "Working with an Embodied and Systemic Approach to Trauma and Tragedy," in *Tragedies and Christian Congregations: The Practical Theology of Trauma*, ed. Megan Warner et al. (Abingdon: Routledge, 2019), 47–63, 51–52.

27. Dori Laub, "Truth and Testimony: The Process and the Struggle," in *Trauma: Explorations in Memory*, ed. Cathy Caruth (Baltimore, MD: Johns Hopkins University Press, 1995), 61–75, 65. Laub is specifically talking about the impossibility and immorality of describing the trauma of Auschwitz—partly because we cannot have any accounts from inside the gas chambers, and partly because it would be "barbaric" to think that we can accurately capture and communicate such a horror.

28. Cathy Caruth, "Recapturing the Past: Introduction," in *Trauma: Explorations in Memory*, ed. Cathy Caruth (Baltimore, MD: Johns Hopkins University Press, 1995), 151–57, 154. As Eric Boynton has it, "all modes of narrative representation and discourse are rejected." See Boynton, "Evil, Trauma, and the Building of Absences," in *Trauma and Transcendence*, ed. Eric Boynton and Peter Capretto (New York: Fordham University Press, 2018), 83–101, 90.

29. Caruth, "Recapturing the Past: Introduction," 154.

30. O'Donnell, *Broken Bodies*, 6.

31. Bessel A. van der Kolk, *The Body Keeps the Score: Brain, Mind, and Body in the Healing of Trauma* (London: Penguin Books, 2015), 86. Van der Kolk's book has

become one of the most influential texts in trauma studies, although it is not without its critics. For example, Natalie Collins suggests that *The Body Keeps the Score* "leaves much to be desired on the feminist front" and is minimal in its engagement with "the endemic nature of patriarchy." See Collins, "Broken or Superpowered? Traumatized People, Toxic Doublethink and the Healing Potential of Evangelical Christian Communities," in *Feminist Trauma Theologies: Body, Scripture & Church in Critical Perspective*, ed. by Karen O'Donnell and Katie Cross (London: SCM Press, 2020), 195–221, 201.

32. O'Donnell, *Broken Bodies*, 7; Rambo, *Spirit and Trauma*, 19.

33. Rambo, *Spirit and Trauma*, 19.

34. Karen O'Donnell, "Eucharist and Trauma: Healing in the B/body," in *Tragedies and Christian Congregations: The Practical Theology of Trauma*, ed. Megan Warner et al. (Abingdon: Routledge, 2019), 182–93, 185.

35. See Chapter 5 for a much fuller discussion of how flesh is to be understood.

36. Will Steffen et al., "Trajectories of the Earth System in the Anthropocene," *PNAS* 115 (2018): 8252–59; Timothy Lenton et al, "Climate Tipping Points—Too Risky to Bet Against," *Nature* 575 (2019): 592–95.

37. As first proposed by Paul J. Crutzen and Eugene F. Stoermer. See Crutzen and Stoermer, "The 'Anthropocene,'" *International Geosphere-Biosphere Programme Global Change Newsletter* (2000), 17–18. However, in 2024, scientists refused to ratify the Anthropocene as a new epoch on the geological timescale. I deal further with the concept of the Anthropocene in Chapter 6.

38. IPCC, "Summary for Policymakers," 21.

39. Judith Lewis Herman, *Trauma and Recovery* (New York: Basic Books, 1992), 35.

40. That said, constructive work in trauma theology may well still be used to formulate pastoral responses. For examples of primarily pastoral theologies of trauma see: Deborah van Deusen Hunsinger, "Bearing the Unbearable: Trauma, Gospel and Pastoral Care," *Theology Today* 68 (2011): 8–25; Lynn Bridgers, "The Resurrected Life: Roman Catholic Resources in Posttraumatic Pastoral Care," *International Journal of Practical Theology* 15 (2011): 38–56; Deanna A. Thompson, *Glimpsing Resurrection: Cancer, Trauma, and Ministry* (Louisville, KY: Westminster John Knox Press, 2018); Megan Warner et al., eds., *Tragedies and Christian Congregations: The Practical Theology of Trauma* (Abingdon: Routledge, 2019); R. Ruard Ganzevoort and Srdjan Sremac, eds., *Trauma and Lived Religion: Transcending the Ordinary* (Cham, Switzerland: Palgrave Macmillan, 2019).

41. Rambo, *Spirit and Trauma*, 79, 168, 170; O'Donnell, *Broken Bodies*, 2, 10, 12–13. As O'Donnell writes, "rather than seeking to mould experiences of trauma to fit with existing doctrines and theologies, these [trauma] theologians begin with the experience of trauma as the 'real' and allow that experience to inform and challenge doctrine." See O'Donnell, *Broken Bodies*, 10. For further examples of theological reassessments in light of trauma see: Serene Jones, "Hope Deferred: Theological Reflections on Reproductive Loss (Infertility, Miscarriage, Stillbirth)," *Modern*

Theology 17 (2001): 227–45; Erin Kidd, "The Violation of God in the Body of the World: A Rahnerian Response to Trauma," *Modern Theology* 35 (2019): 663–82.

42. For example, see: Joanna Collicutt McGrath, "Post-Traumatic Growth and the Origins of Early Christianity," *Mental Health, Religion and Culture* 9 (2006): 291–306; David McLain Carr, *Holy Resilience: The Bible's Traumatic Origins* (New Haven, CT: Yale University Press, 2014); Elizabeth Boase and Christopher G. Frechette, eds., *Bible Through the Lens of Trauma* (Atlanta: SBL Press, 2016); Hilary Jerome Scarsella, "Trauma and Theology," in *Trauma and Transcendence*, ed. Eric Boynton and Peter Capretto (New York: Fordham University Press, 2018), 256–82; Megan Warner, "Trauma Through the Lens of the Bible," in *Tragedies and Christian Congregations: The Practical Theology of Trauma*, ed. Megan Warner et al. (Abingdon: Routledge, 2019), 81–91.

43. Many of the early trauma theologians simply got on with *doing* trauma theology without first explaining their methodology. Several different analytical techniques have been employed: Jennifer Beste uses qualitative data in the form of interviews; Serene Jones adopts autoethnographic, historical, and literary methods; Carrie Doehring employs quantitative statistical data; Marcus Pound engages critical theory; Dirk Lange turns to historical and literary criticism; and Shelly Rambo demonstrates a more explicitly constructive theology. See Karen O'Donnell, "The Voices of the Marys: Towards a Method in Feminist Trauma Theologies," in *Feminist Trauma Theologies: Body, Scripture & Church in Critical Perspective*, ed. Karen O'Donnell and Katie Cross (London: SCM Press, 2020), 3–20, 4, 8; Jennifer Erin Beste, *God and the Victim: Traumatic Intrusions on Grace and Freedom* (Oxford: Oxford University Press, 2007); Serene Jones, *Trauma and Grace: Theology in a Ruptured World*, 2nd Edition (Louisville, KY: Westminster John Knox Press, 2019); Carrie Doehring, *Internal Desecration: Traumatization and Representation of God* (Lanham, MD: University Press of America, 1993); Marcus Pound, *Theology, Psychoanalysis, Trauma* (London: SCM Press, 2007); Dirk G. Lange, *Trauma Recalled: Liturgy, Disruption, and Theology* (Minneapolis: Fortress Press, 2010); Rambo, *Spirit and Trauma*.

44. Karen O'Donnell and Katie Cross, "Introduction," in *Feminist Trauma Theologies: Body, Scripture & Church in Critical Perspective*, ed. Karen O'Donnell and Katie Cross (London: SCM Press, 2020), xix–xxv, xxi–xxii. The editors of *Tragedies and Christian Congregations* also highlight the same pairing of constructive theology and practical theology. See Warner et al., *Tragedies and Christian Congregations*, 11.

45. O'Donnell, "The Voices of the Marys," 5.

46. Jason A. Wyman, "Interpreting the History of the Workgroup on Constructive Theology," *Theology Today* 73 (2017): 312–24, 312.

47. These are features that Elaine Graham identifies in Shelly Rambo's approach to trauma theology. See Graham, "After the Fire, the Voice of God: Speaking of God After Tragedy and Trauma," in *Tragedies and Christian Congregations: The Practical Theology of Trauma*, ed. Megan Warner et al. (Abingdon: Routledge, 2019), 13–27.

48. O'Donnell and Cross, "Introduction," xxii.

49. Graham, "After the Fire, the Voice of God," 20.

50. See Chapter 1 for a much more detailed discussion of the Earth as the subject of trauma.

51. Thompson, *Glimpsing Resurrection*, 4.

52. Jones, *Trauma and Grace*, 22. Shelly Rambo evokes the same sentiment when she quotes from Hans Urs von Balthasar in his introduction to one of the volumes of *Explorations in Theology*: "This is merely a sketchbook: all it tries to do is approach its main object from different angles.... Perhaps some eager soul thirsty for systematics would like to make something out of these fragments.... The author, however, mistrusts such undertakings." See Hans Urs von Balthasar, *Spirit and Institution, Volume 4 of Explorations in Theology*, trans. Edward T. Oakes (San Francisco: Ignatius Press, 1995), 11, quoted in Rambo, *Spirit and Trauma*, 171n78.

53. Shelly Rambo, *Resurrecting Wounds: Living in the Afterlife of Trauma* (Waco, TX: Baylor University Press, 2017), 2.

54. Rambo, *Resurrecting Wounds*, 7. What Rambo encourages is a "reenvisioning" of the resurrection by focusing on the "afterlife of the cross"—and especially the persistence of the wounds on Christ's body. See Rambo, *Resurrecting Wounds*, 1, 6.

55. Rambo, *Spirit and Trauma*, 156.

56. Sallie McFague, "An Ecological Christology: Does Christianity Have It?," in *Christianity and Ecology: Seeking the Well-Being of Earth and Humans*, ed. Dieter T. Hessel and Rosemary Radford Ruether (Cambridge, MA: Harvard University Press, 2000), 29–45, 29.

57. Celia Deane-Drummond, *Eco-Theology* (London: Darton, Longman and Todd, 2008), 99.

58. Serene Jones's work on Christology and trauma serves as an important warning about the limits of systematic thinking. Jones lists five ways in which the passion narrative mirrors the experience of trauma survivors, stating that "I initially believed they would provide me with the substance of the Christology I was looking for, a theology of the cross that would make redemptive sense." But she subsequently realizes that the passion narrative is "simply too narrow to hold all the stories of persons and communities that it should be able to include." Trauma fragments and exceeds any attempt to contain it, even when that container is Christ. As Jones concludes, when faced with trauma, "there is no need for a well-formulated Christology packaged in a singular ending." See Jones, *Trauma and Grace*, 78–79, 97.

59. Niels Henrik Gregersen, "The Cross of Christ in an Evolutionary World," *Dialog* 40 (2001): 192–207, 205. Duncan Reid proposed a very similar idea at a very similar time. See Reid, "Enfleshing the Human: An Earth-Revealing, Earth-Healing Christology," in *Earth Revealing—Earth Healing: Ecology and Christian Theology*, ed. Denis Edwards (Collegeville, MN: The Liturgical Press, 2001), 69–83. The most substantial treatment of deep incarnation to date is Gregersen's edited volume of

2015. See Niels Henrik Gregersen, ed., *Incarnation: On the Scope and Depth of Christology* (Minneapolis: Fortress Press, 2015).

60. Niels Henrik Gregersen, "The Extended Body: The Social Body of Jesus According to Luke," *Dialog* 51 (2012): 234–44, 234. The social nature of Christ's body is also helpfully captured by the distinction in German between *Leib* (a living body with all its interconnections) and *Körper* (a mere physical body or corpse). See Niels Henrik Gregersen, "Christology," in *Climate Change and Systematic Theology: Ecumenical Perspectives*, ed. Michael Northcott and Peter Scott (New York: Routledge, 2014), 33–50, 40. Note, too, how Jesus's understanding of kinship extends far beyond the merely genetic. See Niels Henrik Gregersen, "Cur Deus Caro: Jesus and the Cosmos Story," *Theology and Science* 11 (2013): 370–93, 380.

61. Gregersen, "The Cross of Christ in an Evolutionary World," 205.

62. Niels Henrik Gregersen, "Deep Incarnation: Why Evolutionary Continuity Matters in Christology," *Toronto Journal of Theology* 26 (2010): 173–88, 182.

63. Gregersen, "Deep Incarnation: Why Evolutionary Continuity Matters in Christology," 176.

64. For example: Gregersen, "Cur Deus Caro," 370. I will pursue this line of thinking in Chapter 5.

65. However, for some, deep incarnation does not go far enough. Matthew Eaton criticizes Gregersen for perpetuating a metaphysical anthropocentrism since value is only extended to the nonhuman to the extent that it conforms to the human image. See Eaton, "Enfleshed in Cosmos and Earth: Re-Imagining the Depth of the Incarnation," *Worldviews: Global Religions, Culture, and Ecology* 18 (2014): 230–54.

66. Supralapsarianism moves Christology away from the "sin- and guilt-based ground for the incarnation." See Ronald Cole-Turner, "Incarnation Deep and Wide: A Response to Niels Gregersen," *Theology and Science* 11 (2013): 424–35, 432, 434. As Gregersen writes, "even if Adam and Eve had not sinned, Christ would have come anyway—in order to accomplish a full realization of the goal of creation." See Niels Henrik Gregersen, "Deep Incarnation and Kenosis: In, With, Under, and As: A Response to Ted Peters," *Dialog* 52 (2013): 251–62, 261n20.

67. For example, Mary Daly writes: "The qualities that Christianity idealizes, especially for women, are also those of the victim: sacrificial love, passive acceptance of suffering, humility, meekness, etc. Since these are the qualities idealized in Jesus 'who died for our sins' his functioning as a model reinforces the scapegoat syndrome for women." Daly's concern is that women are encouraged to remain obedient and accept abuse. See Daly, *Beyond God the Father: Towards a Philosophy of Women's Liberation* (Wellingborough: The Women's Press, 1986), 77. Similarly, Delores S. Williams notes how the cross can idolize black women's experience of surrogacy: "Jesus represents the ultimate surrogate figure; he stands in the place of someone else: sinful humankind. Surrogacy, attached to this divine personage, thus takes on an aura of the sacred." See Williams, *Sisters in the Wilderness: The Challenge of Womanist God-Talk* (Maryknoll, NY: Orbis Books, 1993), 143.

68. Kathryn Tanner's suggestion of Christ as the "key" to what God is doing

everywhere is pertinent here. If the entire phenomenon of the incarnation is taken to be the means of redemption—or even, in this case, the means of witness—then there is less risk of abstracting and idolizing the cross. Tanner writes, "serious attention to the incarnation enables one to revise traditional descriptions and explanations of the saving significance of the cross so as to do justice to the criticisms that feminist and womanist theologians rightfully lodge against classical atonement theories." See Tanner, *Christ the Key* (Cambridge: Cambridge University Press, 2010), viii, 247.

69. For example: Colossians 1:15–20.

70. Kwok Pui-lan, "Ecology and Christology," *Feminist Theology* 5 (1989): 113–25, 118.

71. Kwok Pui-lan, "Response to Sallie McFague," in *Christianity and Ecology: Seeking the Well-Being of Earth and Humans*, ed. Dieter T. Hessel and Rosemary Radford Ruether (Cambridge, MA: Harvard University Press, 2000), 47–50, 50. This is also a tension that runs throughout the edited volume *Envisioning the Cosmic Body of Christ*. See Aurica Jax and Saskia Wendel, "Introduction," in *Envisioning the Cosmic Body of Christ: Embodiment, Plurality and Incarnation*, ed. Aurica Jax and Saskia Wendel (Abingdon: Routledge, 2020), 1–4, 1.

72. Stef Craps, *Postcolonial Witnessing: Trauma Out of Bounds* (Basingstoke: Palgrave Macmillan, 2013), 9.

73. Craps, *Postcolonial Witnessing*, 3. See also: Sonya Andermahr, "'Decolonizing Trauma Studies: Trauma and Postcolonialism'—Introduction," *Humanities* 4 (2015): 500–505, 501.

74. Craps, *Postcolonial Witnessing*, 22.

75. Cláudio Carvalhaes, "Colonization, Trauma and Prayers: Towards a Collective Healing," in *Bearing Witness: Intersectional Perspectives on Trauma Theology*, ed. Karen O'Donnell and Katie Cross (London: SCM Press, 2022), 294–310, 295.

76. Christina Elizabeth Sharpe, *In the Wake: On Blackness and Being* (Durham, NC: Duke University Press, 2016), 2. Sharpe draws here on the many meanings of "wake"—including the track left by a ship, a general region of disturbed flow, a watch or vigil for someone who has died, or even a state of wakefulness. See Sharpe, *In the Wake*, 3, 4, 10.

77. Sharpe, *In the Wake*, 15.

78. Craps, *Postcolonial Witnessing*, 25. The phrase "postcolonial traumatic stress disorder" was first used by Tariana Turia, while the term "post-traumatic slavery syndrome" was coined by Alvin Poussaint and Amy Alexander. See Tariana Turia, "Tariana Turia's Speech Notes," *Speech to NZ Psychological Society Conference* (2000), accessed 14 August 2024, http://www.converge.org.nz/pma/tspeech.htm; Alvin F. Poussaint and Amy Alexander, *Lay My Burden Down: Suicide and the Mental Health Crisis Among African Americans* (Boston: Beacon Press, 2000). See also: Joy DeGruy, *Post Traumatic Slave Syndrome: America's Legacy of Enduring Injury and Healing* (Portland, OR: Joy DeGruy Publications, 2005); Shari Renée

Hicks, "A Critical Analysis of Post Traumatic Slave Syndrome: A Multigenerational Legacy of Slavery," unpublished doctoral thesis (California Institute of Integral Studies, 2015), accessed 16 August 2024, https://www.proquest.com/dissertations-theses/critical-analysis-post-traumatic-slave-syndrome/docview/1707689965/se-2?accountid=13042.

79. Kelly Brown Douglas, "Foreword," in *Trauma and Grace: Theology in a Ruptured World*, 2nd Edition, by Serene Jones (Louisville, KY: Westminster John Knox Press, 2019), vii–x, viii.

80. Katie Cross and Karen O'Donnell, "Introduction," in *Bearing Witness: Intersectional Perspectives on Trauma Theology*, ed. Karen O'Donnell and Katie Cross (London: SCM Press, 2022), 1–9, 7.

81. Selina Stone, "Spirit for the Oppressed? Pentecostalism, the Spirit and Black Trauma," in *Bearing Witness: Intersectional Perspectives on Trauma Theology*, ed. Karen O'Donnell and Katie Cross (London: SCM Press, 2022), 58–76, 67.

82. Achille Mbembe, *Necropolitics*, trans. Steve Corcoran (Durham, NC: Duke University Press, 2019), 24.

83. Mbembe, *Necropolitics*, 10.

84. Mbembe, *Necropolitics*, 10.

85. James H. Cone, "Whose Earth Is It, Anyway?," *Cross Currents* 50 (2000): 36–46, 36.

86. Melanie L. Harris, *Ecowomanism: African American Women and Earth-Honoring Faiths*, Ecology and Justice (Maryknoll, NY: Orbis Books, 2017), 3, 6.

87. Mbembe, *Necropolitics*, 93.

88. Some parts of this chapter have previously been published. See Timothy A. Middleton, "Christic Witnessing: A Practical Response to Ecological Trauma," *Practical Theology* 15 (2022): 420–31.

89. Some parts of this chapter have previously been published. See Timothy A. Middleton, "Christology and the Temporal Trauma of the Anthropocene," in *Rethinking Theology in the Anthropocene*, ed. Andreas Krebs (Darmstadt: Wissenschaftliche Buchgesellschaft, 2024), 63–80.

1. The Traumatized Earth

1. Hosea 4:3. All biblical quotations are from the NRSV unless otherwise specified.

2. For instance, Kivatsi Jonathan Kavusa reads these verses as an "expression of trauma in the natural world." See Kavusa, "Social Disorder and the Trauma of the Earth Community: Reading Hosea 4:1–3 in Light of Today's Crises," *Old Testament Essays* 3 (2016): 481–501, 499.

3. Jeremiah 4:23, 28.

4. Jeremiah 12:4.

5. Joel 1:10, 20. Further examples can be found in Isaiah 24:1–20; Isaiah 33:7–9; Jeremiah 12:7–13; Jeremiah 23:9–12; Ezekiel 19:7; Amos 1:2.

6. Richard Bauckham, "The Story of the Earth According to Paul: Romans 8:18–23," *Review & Expositor* 108 (2011): 91–97, 94.

7. Katherine Murphey Hayes, *The Earth Mourns: Prophetic Metaphor and Oral Aesthetic* (Leiden: Brill, 2002), 243n23.

8. See Hosea 4:2.

9. Christoph Uehlinger, "The Cry of the Earth? Biblical Perspectives on Ecology and Violence," in *Ecology and Poverty: Cry of the Earth, Cry of the Poor*, ed. Leonardo Boff and Virgil Elizondo (London: SCM Press, 1995), 41–57, 46.

10. Terence E. Fretheim, *God and World in the Old Testament: A Relational Theology of Creation* (Nashville, TN: Abingdon Press, 2005), 267–68.

11. Psalm 19:2; Psalm 19:2; Psalm 65:13; Psalm 89:12; Psalm 93:3; Psalm 96:11; Psalm 96:12; Isaiah 35:1; Isaiah 42:11; Isaiah 55:12; Isaiah 66:23.

12. Psalm 148:3–10.

13. Luke 19:40.

14. David G. Horrell and Dominic Coad, "'The Stones Would Cry Out' (Luke 19:40): A Lukan Contribution to a Hermeneutics of Creation's Praise," *Scottish Journal of Theology* 64 (2011): 29–44, 42. Talking stones might sound far-fetched to Western ears, but it is deeply reminiscent of Indigenous wisdom when George "Tink" Tinker says that "rocks talk and have what we must call consciousness." In Tinker's words, a stone that shouts out is "a way of seeing the world and living our lives." See Tinker, "The Stones Shall Cry Out: Consciousness, Rocks, and Indians," *Wicazo Sa Review* 19 (2004): 105–25, 106, 122.

15. Romans 8:22.

16. Bauckham, "The Story of the Earth According to Paul," 93. Furthermore, consensus tends to be that Paul is referring to all of the nonhuman creation, whether animate or inanimate. See Cherryl Hunt et al., "An Environmental Mantra? Ecological Interest in Romans 8: 19–23 and a Modest Proposal for Its Narrative Interpretation," *Journal of Theological Studies* 59 (2008): 546–79, 558.

17. Sigve K. Tonstad, *The Letter to the Romans: Paul Among the Ecologists* (Sheffield: Sheffield Phoenix Press, 2016), 243.

18. As some commentators put it, "the groaning of Earth in Romans 8 seems to be a silenced voice finally heard in the face of death. Perhaps the same voice is to be heard in the ecological crisis of our day." But in the ecological crisis of our day, the Earth is groaning not because it is *en route* to some future improvement, but because it is mourning what is being irretrievably destroyed. See Norman C. Habel with the Earth Bible Team, "Where Is the Voice of Earth in Wisdom Literature?," in *The Earth Story in Wisdom Traditions*, ed. Norman C. Habel and Shirley Wurst (Sheffield: Sheffield Academic Press, 2001), 23–34, 33.

19. See Richard Bauckham, "Joining Creation's Praise of God," *Ecotheology* 7 (2002): 45–59, 47. In the case of creation's praise, there are some other possible explanations in the literature: that creation's praise is meant to inspire human praise, that nonhuman entities should be read as allegories for human entities, that this is a foretaste of the eschaton where nonhumans will have the literal ability to praise

God, or that this is to contrast the Israelite faith with other Ancient Near Eastern religions by emphasizing that the sun and moon are givers of praise to God rather than objects of worship in themselves. See Fretheim, *God and World in the Old Testament*, 252–54; Elizabeth Johnson, "Animals' Praise of God," *Interpretation: A Journal of Bible and Theology* 73 (2019): 259–71, 262–63.

20. Habel with the Earth Bible Team, "Where Is the Voice of Earth in Wisdom Literature?," 23.

21. Habel with the Earth Bible Team, "Where Is the Voice of Earth in Wisdom Literature?," 23.

22. Earth Bible Team, "The Voice of Earth: More than Metaphor?," in *The Earth Story in the Psalms and the Prophets*, ed. Norman C. Habel (Sheffield: Sheffield Academic Press, 2001), 23–28, 24.

23. Earth Bible Team, "The Voice of Earth," 24.

24. "If my land has cried out against me, and its furrows have wept together; if I have eaten its yield without payment, and caused the death of its owners; let thorns grow instead of wheat, and foul weeds instead of barley." See Job 31:38–40.

25. Pope Francis, *Encyclical Letter Laudato Si' of the Holy Father Francis: On Care for Our Common Home* (Vatican City: Vatican Press, 2015), §92.

26. Francis, *Laudato Si'*, §1.

27. Francis, *Laudato Si'*, §2. The encyclical then links this cry to the groaning of creation in Romans 8:22. However, as I explained above, I am not convinced that this is the best scriptural analogue for ecological trauma, or our current ecological crisis in general, because of Paul's teleological insinuations.

28. Francis, *Laudato Si'*, §53.

29. Francis, *Laudato Si'*, §49. This phrase about "the cry of the Earth and the cry of the poor" echoes Leonardo Boff's book of the same name. Francis never cites Boff, although Boff's work is undoubtedly in the background here. See Boff, *Cry of the Earth, Cry of the Poor*, trans. Phillip Berryman (Maryknoll, NY: Orbis Books, 1997).

30. Bruno Latour, "The Immense Cry Channeled by Pope Francis," *Environmental Humanities* 2 (2016): 251–55, 253. Latour also deals with the attribution of agency and subjectivity to the Earth in his Gifford lectures, *Facing Gaia*. He is especially struck by the fact that, amid our contemporary ecological crisis, we are the ones standing by as "witless objects" while the planet takes the role of "active subject." See Latour, *Facing Gaia: Eight Lectures on the New Climatic Regime*, trans. Catherine Porter (Cambridge: Polity Press, 2017), 73.

31. Latour, "The Immense Cry Channeled by Pope Francis," 255.

32. Francis, *Laudato Si'*, §90.

33. Francis, *Laudato Si'*, §246.

34. Roger D. Sorrell, *St. Francis of Assisi and Nature: Tradition and Innovation in Western Christian Attitudes Toward the Environment* (Oxford: Oxford University Press, 2009), 128.

35. Bron Raymond Taylor refers to these diverse traditions as "dark green religion," noting that they are united and defined by the view that "nature is sacred,

has intrinsic value, and is therefore due reverent care." But he is also clear that dark green religion is an "eclectic bricolage," that is, "an amalgamation of bits and pieces of a wide array of ideas and practices, drawn from diverse cultural systems, religious traditions, and political ideologies." See Taylor, *Dark Green Religion: Nature Spirituality and the Planetary Future* (Berkeley: University of California Press, 2010), 10, 14.

36. Plato, *Plato: Timaeus and Critias*, ed. and trans. Alfred Edward Taylor (Abingdon: Routledge, 2013), 27, 30b.

37. Douglas Hedley, "Sophia and the World Soul," in *The Cambridge Companion to Christianity and the Environment*, ed. Alexander J. B. Hampton and Douglas Hedley (Cambridge: Cambridge University Press, 2022), 286–302.

38. Hedley, "Sophia and the World Soul," 292.

39. J. N. D. Kelly, *Early Christian Doctrines* (London: Adam and Charles Black, 1975), 285n21.

40. It might make intuitive sense to position the second person of the Trinity as the subject of ecological trauma because it is already widely accepted that God suffers in Christ. In the incarnation and the crucifixion, Christ suffers and dies— traumatically.

41. Theodore Roszak, *The Voice of the Earth: An Exploration of Ecopsychology* (New York: Simon & Schuster, 1992), 140.

42. Philosophical advocates of panpsychism include William James, Alfred North Whitehead, Bertrand Russell, Galen Strawson, Thomas Nagel, and David Chalmers.

43. This is the argument made by Thomas Nagel. See Nagel, *Mind and Cosmos: Why the Materialist Neo-Darwinian Conception of Nature Is Almost Certainly False* (New York: Oxford University Press, 2012).

44. Joanna Leidenhag, *Minding Creation: Theological Panpsychism and the Doctrine of Creation*, T & T Clark Studies in Systematic Theology (London: T & T Clark, 2020); Joanna Leidenhag, "Panpsychism and God," *Philosophy Compass* (2022): 1–11.

45. Leidenhag, *Minding Creation*, 168.

46. Leidenhag, *Minding Creation*, 144, 149.

47. Leidenhag, *Minding Creation*, 5, 107.

48. Leidenhag, *Minding Creation*, 62.

49. James Lovelock, *Gaia: A New Look at Life on Earth* (Oxford: Oxford University Press, 2000), 9.

50. Lovelock, *Gaia: A New Look at Life on Earth*, 138. Although biologists were quick to point out that Gaia cannot be formally alive if it cannot reproduce.

51. Stephan Harding, "Animate Earth," in *Earthy Realism: The Meaning of Gaia*, ed. Mary Midgley (Exeter: Imprint Academic, 2007), 23–29, 25. As such, for Stephan Harding, "a gaping construction site, for example, or a clear-cut mountainside, may communicate the genuine, objective suffering of the Earth." See Harding, *Animate Earth: Science, Intuition and Gaia* (Totnes: Green Books, 2006), 21.

52. Louis Heyse-Moore, "Does Gaia Experience Trauma?," *European Journal of Ecopsychology* 7 (2022): 75–99, 75. Mark I. Wallace reaches a very similar conclusion, suggesting that "when we assail Gaia's ecosystemic balance . . . we are causing Earth . . . to suffer harm, to feel pain, and to undergo trauma." See Wallace, "Even Rocks Are Alive: Christian Animist Disruptions of the Species Divide," in *Taking a Deep Breath for the Story to Begin. . .*, ed. Ernst M. Conradie and Pan-Chiu Lai (Cape Town: AOSIS, 2021), 241–58, 253.

53. Heyse-Moore, "Does Gaia Experience Trauma?," 84.

54. John Maynard Smith, quoted in Mary-Jane Rubenstein, *Pantheologies: Gods, Worlds, Monsters* (New York: Columbia University Press, 2018), 119.

55. See Rubenstein, *Pantheologies*, 120; Harding, *Animate Earth*, 28.

56. Lovelock, *Gaia: A New Look at Life on Earth*, ix-x.

57. James Lovelock, *Gaia: The Practical Science of Planetary Medicine* (London: Gaia, 1991), 31.

58. Lovelock, *Gaia: The Practical Science of Planetary Medicine*, 11. The notion of the Newtonian "machine" or Richard Dawkins's "selfish" gene would be good examples.

59. Lovelock, *Gaia: A New Look at Life on Earth*, xiii.

60. James Lovelock, *The Revenge of Gaia: Why the Earth Is Fighting Back—and How We Can Still Save Humanity* (New York: Penguin, 2007), 135–39.

61. Lovelock, *The Revenge of Gaia*, 147.

62. Mary Midgley, *Gaia: The Next Big Idea* (London: Demos, 2001), 11.

63. Ecopsychologists often come from a tradition of Jungian psychology. As Jung writes, "at times I feel as if I am spread out over the landscape and inside things, and am myself living in every tree, in the splashing of the waves, in the clouds and the animals that come and go, in the procession of the seasons." See Carl Gustav Jung, *Memories, Dreams, Reflections* (New York: Random House, 1961), 225, quoted in Meredith Sabini, ed., *The Earth Has A Soul: C. G. Jung on Nature, Technology & Modern Life* (Berkeley, CA: North Atlantic Books, 2016), 14.

64. Theodore Roszak, *Person/Planet* (London: Granada, 1981), 26.

65. Andy Fisher, *Radical Ecopsychology: Psychology in the Service of Life*, 2nd Edition (Albany: State University of New York Press, 2013), 122.

66. Roszak, *Person/Planet*, 78. Roszak goes on to say that "the cry of personal pain which [we utter] . . . is the planet's own cry for rescue"—a statement that seems to anticipate the intertwined cry of the Earth and cry of the poor in *Laudato Si'*. See Roszak, *Person/Planet*, 79.

67. Joanna Macy, *World as Lover, World as Self* (Berkeley, CA: Parallax Press, 2007), 28.

68. There is some overlap here with theories of extended and embodied mind. For example, see: Thomas Fuchs, *Ecology of the Brain: The Phenomenology and Biology of the Embodied Mind* (Oxford: Oxford University Press, 2018).

69. Roszak, *The Voice of the Earth*, 320.

70. Graham Harvey, *Animism: Respecting the Living World* (New York: Columbia University Press, 2006), 11.

71. According to Shawn Sanford Beck, Christian animism "is simply what happens when a committed Christian engages the world and each creature as alive, sentient, and related, rather than soul-less and ontologically inferior." For Beck, such a position is "well within the boundaries of doctrinal orthodoxy" because it need not necessarily be linked with either pantheism or polytheism. See Beck, *Christian Animism* (Winchester: Christian Alternative, 2015), 8, 15.

72. Mark I. Wallace, "Elegy for a Lost World," in *Post-Traumatic Public Theology*, ed. Stephanie N. Arel and Shelly Rambo (Cham, Switzerland: Palgrave Macmillan, 2016), 138.

73. Mark I. Wallace, *When God Was a Bird* (New York: Fordham University Press, 2019), 144.

74. Wallace, "Elegy for a Lost World," 138. Note that there is an important difference between asserting Earth's animacy and inviting us to "re-imagine" Earth's animacy. An emphasis on reimagination is not so far from the anthropomorphic relationality I describe below.

75. As well as the examples given above, Wallace references the story of Cain and Abel. When Cain kills Abel (Genesis 4:9–12), the Earth does not remain passive, but "cries out," "opens its mouth," "swallows," and "curses." The Earth has a vitality and an emotional life such that it experiences the catastrophic loss of Abel's death. "Aggrieved and bereaved, Earth weeps or shouts at the terrible harm Cain has done to his brother and, it seems, to the land as well." See Wallace, "Even Rocks Are Alive," 253n860; Wallace, "Elegy for a Lost World," 139.

76. Mark I. Wallace, "Christian Animism, Green Spirit Theology, and the Global Crisis Today," *Journal of Reformed Theology* 6 (2012): 216–33, 216. Michael S. Northcott offers a slightly different account, defining animism as "what Christians call the *logos*, or divine spirit, that animates all living beings including plants, animals and the winds that circle the Earth." See Northcott, "Religious Traditions and Ecological Knowledge," in *The Cambridge Companion to Christianity and the Environment*, ed. Alexander J. B. Hampton and Douglas Hedley (Cambridge: Cambridge University Press, 2022), 231–46, 231.

77. Wallace, *When God Was a Bird*, 15. Wallace's pneumatological focus means that his Christology is often somewhat muted. Yet he does appeal to Chalcedonian Christology for "theological syntax" that can hold open the "nonoppositional dualism" that he has in mind here; the Spirit relates to the world in an analogous manner to the coinherence of the divine and the human in Christ.

78. Wallace, "Christian Animism, Green Spirit Theology, and the Global Crisis Today," 227.

79. Wallace, "Christian Animism, Green Spirit Theology, and the Global Crisis Today," 228.

80. Wallace, "Elegy for a Lost World," 141.

81. Roszak, *Person/Planet*, 73.

82. Lynn White, "The Historical Roots of Our Ecologic Crisis," *Science* 155 (1967): 1203–7, 1205.

83. Pat Zukeran and Robert Sirico, quoted in Mark I. Wallace, "The Lord God Bird: Avian Divinity, Neo-Animism, and the Renewal of Christianity at the End of the World," in *Encountering Earth: Thinking Theologically With a More-Than-Human World*, ed. Trevor Bechtel et al. (Eugene, OR: Cascade Books, 2018), 210–26, 220.

84. Rubenstein, *Pantheologies*, xx.

85. Tinker, "The Stones Shall Cry Out," 122.

86. Roszak, *The Voice of the Earth*, 95, 97.

87. Lovelock, *Gaia: A New Look at Life on Earth*, xiii; Lovelock, *Gaia: The Practical Science of Planetary Medicine*, 31.

88. Jane Bennett, *Vibrant Matter: A Political Ecology of Things* (Durham, NC: Duke University Press, 2010), xviii, 18. Wallace also points out Bennett's nervousness. See Wallace, "Even Rocks Are Alive," 247.

89. Edward Burnett Tylor, *Religion in Primitive Culture* (New York: Harper & Row, [1871] 1958), 5.

90. Leading thinkers here include Eduardo Viveiros de Castro, Philippe Descola, and Bruno Latour. For one example in an ecological context, see: Eduardo Kohn, *How Forests Think: Toward an Anthropology Beyond the Human* (Berkeley: University of California Press, 2013).

91. Vine Deloria Jr., "If You Think About It, You Will See That It Is True," in *Spirit & Reason: The Vine Deloria, Jr., Reader*, ed. Barbara Deloria et al. (Golden, CO: Fulcrum, 1999), 49.

92. Walking Buffalo, quoted in Tinker, "The Stones Shall Cry Out," 105.

93. Mari Joerstad, *The Hebrew Bible and Environmental Ethics: Humans, Non-Humans, and the Living Landscape* (Cambridge: Cambridge University Press, 2019), 7.

94. Brian J. Walsh et al., "Trees, Forestry, and the Responsiveness of Creation," *Cross Currents* 44 (1994): 149–62, 152.

95. Joerstad, *The Hebrew Bible and Environmental Ethics*, 2.

96. Wallace, *When God Was a Bird*, 165.

97. Wallace, "Even Rocks Are Alive," 255. Moreover, some scholars suggest that even when animist ontologies are operational this is no guarantee of more sustainable behaviors. See Aike P. Rots and Nhung Lu Rots, "When Gods Drown in Plastic: Vietnamese Whale Worship, Environmental Crises, and the Problem of Animism," *Environmental Humanities* 15 (2023): 8–29.

98. Lovelock, *The Revenge of Gaia*, 139.

99. Roszak is quick to follow this up with the observation that, for Indigenous peoples, animism is not "just" a metaphor but "how they truly see the world." See Roszak, *The Voice of the Earth*, 82.

100. Rubenstein, *Pantheologies*, 93.

101. Wallace, "Elegy for a Lost World," 138; Wallace, *When God Was a Bird*, 145. Understanding the Earth as a suffering subject, says Wallace, allows us to: resituate

ourselves with respect to "the wider personhood of Earth itself"; reimagine ourselves as part of something much bigger than our own existences; recognize ourselves as "coparticipants in the web of life"; and motivate ourselves to live in greater harmony with this animate Earth. See Wallace, *When God Was a Bird*, 151–52.

102. Francis, *Laudato Si'*, §246.

103. Martin Buber, *I and Thou*, trans. Ronald Gregor Smith (Edinburgh: T & T Clark, [1923] 2004), 14–15.

104. H. Paul Santmire, "Behold the Lilies: Martin Buber and the Contemplation of Nature," *Dialog* 57 (2018): 18–22, 19. See also: H. Paul Santmire, "I-Thou, I-It, and I-Ens," *The Journal of Religion* 48 (1968): 260–73, 264. In a postscript to the last edition of *I and Thou* Buber directly addresses how the I-Thou relationship pertains to natural entities. He describes how animals can be understood to be on the "threshold of mutuality" while plants are still "prethreshold." Yet, in all cases, some sort of reciprocity is still possible. See Buber, *I and Thou*, 94–95.

105. Santmire, "Behold the Lilies," 20.

106. Paul Ricœur, *The Rule of Metaphor: The Creation of Meaning in Language*, trans. Robert Czerny et al. (London: Routledge, 2003), 4; George Lakoff and Mark Johnson, *Metaphors We Live By* (Chicago: University of Chicago Press, 1980), 3. As Elizabeth Johnson notes, "far from being a simple rhetorical flourish, metaphor has a cognitive function insofar as it changes perceptual awareness." See Johnson, "Animals' Praise of God," 263.

107. Earth Bible Team, "The Voice of Earth," 28.

108. Joerstad, *The Hebrew Bible and Environmental Ethics*, 41. Richard Bauckham argues that if we learned to partake in creation's praise, then we could begin to tackle urban isolation and our instrumentalization of nature. Anthropomorphic metaphor not only helps us to understand the natural world, but also changes our relationship to it. See Bauckham, "Joining Creation's Praise of God," 48.

109. Ludwig Feuerbach, *The Essence of Christianity*, trans. George Eliot (Mineola, NY: Dover Publications, [1881] 2008). However, Feuerbach never addresses why human beings are so psychologically inclined to projection, and he does not consider the role that projection plays in the formation of relationships with the world around us.

110. As Trevor Bechtel, Matthew Eaton, and Timothy Harvie put it, "the more-than-human world is other than a passive object to be interpreted, and instead an active plurality of voices that reaches out and expresses itself within the broader Earth community." See Bechtel et al., "Introduction," in *Encountering Earth: Thinking Theologically With a More-Than-Human World*, ed. Trevor Bechtel et al. (Eugene, OR: Cascade Books, 2018), 1–13, 4.

111. Bechtel et al., "Introduction," 5.

112. Bechtel et al., "Introduction," 6. See also: Hans Jonas, "Epilogue: The Outcry of Mute Things," in *Mortality and Morality: A Search for the Good After Auschwitz* (Evanston, IL: Northwestern University Press, 1996), 198–202, 202.

113. Hayes, *The Earth Mourns*, 235.

114. Richard Bauckham also sounds the same warning. "It is not our vocation to absorb the whole created world into our own human life," he says. See Bauckham, "Joining Creation's Praise of God," 52.

115. Joerstad, *The Hebrew Bible and Environmental Ethics*, 16, 38, 132, 142. However, anthropomorphism can also be understood as part of our primary mode of relationship to other entities, rather than a subsequent step that is imposed on what we have previously decided is impersonal.

116. Tinker, "The Stones Shall Cry Out," 107. Mari Joerstad notes how human experience can only ever be an approximate guide for understanding nonhuman entities, but that it is better than no guide at all. See Joerstad, *The Hebrew Bible and Environmental Ethics*, 133.

117. Buber, *I and Thou*, 15. Note, however, that if we are hoping for a sympathetic relationship with the tree, then we are relying on the assumption that humans treat other humans with appropriate compassion and respect. Tragically, this is not always true. Indeed, many forms of human oppression operate by dehumanizing and objectifying other human beings. Nevertheless, anthropomorphism can hopefully expand, rather than diminish, our field of compassion.

118. Bennett, *Vibrant Matter*, 120.

119. Fisher, *Radical Ecopsychology*, 135.

120. Gabriella Airenti, "The Development of Anthropomorphism in Interaction: Intersubjectivity, Imagination, and Theory of Mind," *Frontiers in Psychology* 9 (2018): 1–13, 2.

121. Airenti, "The Development of Anthropomorphism in Interaction," 8.

122. Airenti, "The Development of Anthropomorphism in Interaction," 8.

123. Airenti, "The Development of Anthropomorphism in Interaction," 8.

124. There is significant overlap here between my argument and the adoption of anthropomorphic ways of thinking by animal ethologists. Celia Deane-Drummond summarizes this work from a theological perspective, noting that "to assume other creatures do not have such [emotional] states may be just as problematic as assuming they do." Both positions claim to know something that cannot ultimately be determined. Anthropomorphism is therefore justified as a "heuristic tool to understand how other animals think and act more clearly." See Deane-Drummond, *The Wisdom of the Liminal: Evolution and Other Animals in Human Becoming* (Grand Rapids, MI: Eerdmans, 2014), 42, 110. Moreover, as Marc Bekoff explains, anthropomorphism is a mode of relationship rather than an assertion of equivalence: "The way human beings describe and explain the behavior of other animals is limited by the language they use to talk about things in general. By engaging in anthropomorphism—using human terms to explain animals' emotions or feelings—humans make other animals' worlds accessible to themselves. But this is not to say that other animals are happy or sad in the same ways in which humans (or even other conspecifics) are happy or sad." See Bekoff, "Animal Emotions: Exploring Passionate Natures," *BioScience* 50 (2000): 861–70, 867. Bekoff's approach is also not

dissimilar from Gordon M. Burghardt's notion of critical anthropomorphism. See Burghardt, "Animal Awareness: Current Perceptions and Historical Perspective," *American Psychologist* 40 (1985): 905–19, 917. Nevertheless, my proposal goes further than the animal ethologists in that I am also interested in anthropomorphic relations with ecological entities beyond animals.

125. Macy, *World as Lover, World as Self*, 17.

126. Lauren Woolbright, "Wounded Planet, Wounded People: The Possibility of Ecological Trauma," unpublished master's thesis (Clemson University, 2011), 35, accessed 28 October 2018, https://www.proquest.com/docview/881256145.

2. Trauma in Ecotheology

1. Catherine Keller, *Political Theology of the Earth: Our Planetary Emergency and the Struggle for a New Public* (New York: Columbia University Press, 2018), 93.

2. Clive Hamilton, "A Letter From Canberra: The Apocalyptic Fires in Australia Signal Another Future," *Sierra: The Magazine of the Sierra Club*, January 2020, https://www.sierraclub.org/sierra/letter-canberra.

3. Matthias M. Boer et al., "Unprecedented Burn Area of Australian Mega Forest Fires," *Nature Climate Change* 10 (2020): 170–72.

4. For example, see: Paul Komesaroff and Ian Kerridge, "A Continent Aflame: Ethical Lessons From the Australian Bushfire Disaster," *Journal of Bioethical Inquiry* 17 (2020): 11–14.

5. Hamilton, "A Letter From Canberra."

6. Hamilton, "A Letter From Canberra."

7. However, the language of crisis can become dangerous when it is used to justify states of exception, such as the violation of moral norms or the practice of problematic politics. See Kyle Whyte, "Against Crisis Epistemology," in *Routledge Handbook of Critical Indigenous Studies*, ed. Brendan Hokowhitu et al. (London: Routledge, 2020), 52–64.

8. Extinction Rebellion, *This Is Not a Drill: The Extinction Rebellion Handbook* (London: Penguin Books, 2019), 1–2.

9. Damian Carrington, "Why the Guardian Is Changing the Language It Uses About the Environment," *The Guardian*, May 2019, accessed 11 April 2024, https://www.theguardian.com/environment/2019/may/17/why-the-guardian-is-changing-the-language-it-uses-about-the-environment.

10. Rosamund Ions and Kate Wild, "The Language of Climate Change and Environmental Sustainability," *OED* (2021), accessed 11 April 2024, https://www.oed.com/discover/the-language-of-climate-change/?tl=true.

11. Katharine Hayhoe, "When Facts Are Not Enough," *Science* 360 (2018): 943.

12. Timothy Morton, *Hyperobjects: Philosophy and Ecology After the End of the World* (Minneapolis: University of Minnesota Press, 2013), 8.

13. Morton, *Hyperobjects*, 8.

14. Morton, *Hyperobjects*, 8–9.

15. Morton, *Hyperobjects*, 8.

16. Zhiwa Woodbury, "Climate Trauma: Toward a New Taxonomy of Trauma," *Ecopsychology* 11 (2019): 1–8, 3.

17. Woodbury, "Climate Trauma," 6.

18. Although altered behaviors are undoubtedly crucial, the empirical success of the language of trauma to motivate change is not my primary focus here. Rather, I aim to take this linguistic shift seriously by explaining what ecological trauma entails and how it might be conceived theologically.

19. As Christopher Southgate puts it, "a whole strategy of being alive on the planet, a whole quality of living experience is lost when any organism becomes extinct." See Southgate, *The Groaning of Creation: God, Evolution, and the Problem of Evil* (Louisville, KY: Westminster John Knox Press, 2008), 9.

20. See Celia Deane-Drummond, *A Primer in Ecotheology: Theology for a Fragile Earth* (Eugene, OR: Cascade Books, 2017), 1, 10–11.

21. Pope Francis, *Encyclical Letter Laudato Si' of the Holy Father Francis: On Care for Our Common Home* (Vatican City: Vatican Press, 2015), §13, 25, 48, 110, 122, 145, 146, 198, 206, 231.

22. See Holmes Rolston III, "Disvalues in Nature," *The Monist* (1992): 250–78; Wayne Ouderkirk, "Can Nature Be Evil? Rolston, Disvalue, and Theodicy," *Environmental Ethics* 21 (1999): 135–50; Christopher Southgate, "Divine Glory in a Darwinian World," *Zygon* 49 (2014): 784–807, 785; Bethany N. Sollereder, *God, Evolution, and Animal Suffering: Theodicy Without a Fall* (London: Routledge, 2019), 5. Note that the dispassionate term "disvalue" is employed quite deliberately in several of these cases because the authors are restricting the discussion to natural processes and they do not hold the natural world to be fallen, so they want to avoid the more emotive language of natural evil.

23. Francis, *Laudato Si'*, §15, 63, 101, 119, 201, 209.

24. Michael S. Northcott, *The Environment and Christian Ethics* (Cambridge: Cambridge University Press, 1996), 162.

25. Ecumenical Patriarch Bartholomew of Constantinople was one of the first proponents of the idea of ecological sin, declaring that "to commit a crime against the natural world is a sin." See John Chryssavgis, *Cosmic Grace, Humble Prayer: The Ecological Vision of the Green Patriarch* (Grand Rapids, MI: Eerdmans, 2003), 217–22. For another (quite different) take on ecological sin see: Sallie McFague, *The Body of God: An Ecological Theology* (London: SCM Press, 1993), 113.

26. Northcott, *The Environment and Christian Ethics*, 162.

27. Lynn White, "The Historical Roots of Our Ecologic Crisis," *Science* 155 (1967): 1203–7.

28. Northcott, *The Environment and Christian Ethics*, 232.

29. Komesaroff and Kerridge, "A Continent Aflame," 12.

30. Markus Vogt, *Christian Environmental Ethics: Foundations and Central Challenges*, trans. Gary Slater (Paderborn: Brill Schöningh, 2023), 3.

31. Christopher Southgate, *Theology in a Suffering World: Glory and Longing*

(Cambridge: Cambridge University Press, 2018), 1. Some cosmic fall theologies hold that human sin has had a retroactive influence on the prior suffering of creation. But Southgate (and others) find this chronologically implausible.

32. Michael Lloyd, "The Fallenness of Nature: Three Nonhuman Suspects," in *Finding Ourselves After Darwin: Conversations on the Image of God, Original Sin, and the Problem of Evil*, ed. Stanley P. Rosenberg (Grand Rapids, MI: Baker Academic, 2018), 262–79; Rolston, "Disvalues in Nature"; Nancey Murphy, "Science and the Problem of Evil: Suffering as a By-Product of a Finely Turned Cosmos," in *Physics and Cosmology: Scientific Perspectives on the Problem of Natural Evil*, ed. Robert John Russell and William R. Stoeger (Vatican City: Vatican Observatory Publications, 2007), 131–52; Jay B. McDaniel, *Of God and Pelicans: A Theology of Reverence for Life* (Louisville, KY: Westminster John Knox Press, 1989); Thomas Jay Oord, "An Open Theology Doctrine of Creation and Solution to the Problem of Evil," in *Creation Made Free: Open Theology Engaging Science*, ed. Thomas Jay Oord (Eugene, OR: Wipf & Stock, 2009), 28–52. See also: Southgate, *The Groaning of Creation*; Michael J. Murray, *Nature Red in Tooth and Claw: Theism and the Problem of Animal Suffering* (Oxford: Oxford University Press, 2008); Nicola Hoggard Creegan, *Animal Suffering and the Problem of Evil* (New York: Oxford University Press, 2013); Sollereder, *God, Evolution, and Animal Suffering*.

33. Philippe Descola, *Beyond Nature and Culture*, trans. Janet Lloyd (Chicago: University of Chicago Press, 2013), 57–88.

34. Bruno Latour, *We Have Never Been Modern*, trans. Catherine Porter (Cambridge, MA: Harvard University Press, 1993), 2.

35. Latour, *We Have Never Been Modern*, 7.

36. Roger Abbott and Bob White go one step further. They argue, not just that moral evil and natural evil are deeply intertwined, but that there is no such thing as genuinely natural evil. For example, they propose that several human factors—including colonial oppression, government corruption, poor infrastructure development, widespread deforestation, and rapid urbanization—were ultimately responsible for turning the 2010 Haitian earthquake into an "unnatural disaster." As they write, "we conclude that what happened in Haiti . . . to make it the disaster it continues to be, was not natural, but something that was the consequence of accumulated human evil." See Abbott and White, "Haiti—An Unnatural Disaster," *Ethics in Brief* 18 (2013): 1–4, 2–3. In his subsequent book, White goes on to suggest that "it is almost always the decisions and actions of humans which turn an otherwise beneficial natural process into a disaster." But the word "almost" is crucial; natural evil can never be eliminated entirely. Furthermore, White reinforces the view that moral evil and natural evil compete in a zero-sum game, when in fact the two agencies are regularly interwoven. See Robert S. White, *Who Is to Blame? Disasters, Nature and Acts of God* (Oxford: Monarch Books, 2014), 10.

37. Celia Deane-Drummond, *Eco-Theology* (London: Darton, Longman and Todd, 2008), 116.

38. Serene Jones, *Trauma and Grace: Theology in a Ruptured World*, 2nd Edition (Louisville, KY: Westminster John Knox Press, 2019), 154.

39. Jones, *Trauma and Grace*, 155.

40. Shelly Rambo, *Spirit and Trauma: A Theology of Remaining* (Louisville, KY: Westminster John Knox Press, 2010), 40, 111–41, 151–52.

41. Roger S. Gottlieb, *Morality and the Environmental Crisis* (Cambridge: Cambridge University Press, 2019), 190.

42. Vogt, *Christian Environmental Ethics*, xxiii.

43. Clive Hamilton, "Crimes Against Nature: The Banality of Ethics in the Anthropocene," *ABC Religion & Ethics* (2015), accessed 6 May 2019, https://www.abc.net.au/religion/crimes-against-nature-the-banality-of-ethics-in-the-anthropocene/10098110. The phrase "banality of ethics" echoes Hannah Arendt's famous comments about the banality of evil during the Nazi regime, and there are similarities in terms of the failure of ethical categories—especially an ethics that relies on moral conscience—in the face of something so overwhelming. As Arendt writes of the Third Reich: "When we were first confronted with it, it seemed, not only to me but to many others, to transcend all moral categories. . . . We shall not be able to become reconciled to it, to come to terms with it." See Arendt, *Responsibility and Judgment*, ed. Jerome Kohn (New York: Schocken Books, 2003), 55.

44. Hamilton, "Crimes Against Nature," para. 20. Dale Jamieson pursues a similar line of argument. He writes, for example, "that the challenges that climate change presents go beyond the resources of commonsense morality." Since the impacts of climate change are distributed widely across time and space it becomes nigh on impossible to pinpoint responsibility, and this becomes a challenge to morality itself. See Jamieson, *Reason in a Dark Time: Why the Struggle Against Climate Change Failed—and What It Means for Our Future* (Oxford: Oxford University Press, 2014), 6. The very title of Jamieson's book indicates the failure of reason.

45. Clive Hamilton et al., "Thinking the Anthropocene," in *The Anthropocene and the Global Environment Crisis—Rethinking Modernity in a New Epoch*, ed. Clive Hamilton et al. (Abingdon: Routledge, 2015), 1–13, 8.

46. Cathy Caruth, "Recapturing the Past: Introduction," in *Trauma: Explorations in Memory*, ed. Cathy Caruth (Baltimore, MD: Johns Hopkins University Press, 1995), 151–57, 153. Yet, as Eric Boynton and Peter Capretto note, there is a danger that this focus on the non-intelligibility and non-assimilability of trauma verges on obscurantism. See Boynton and Capretto, "Introduction," in *Trauma and Transcendence*, ed. Eric Boynton and Peter Capretto (New York: Fordham University Press, 2018), 1–14. If trauma is completely and utterly unintelligible, then labeling a phenomenon as traumatic adds nothing to our understanding because no generalizations or connections can be made. On this reading, trauma becomes a synonym for all aporias and lacunae in previous discourse without ever adding anything meaningful or conceptual. Mary-Jane Rubenstein, in her afterword to Boynton and Capretto's book, summarizes trauma's twin problem: "Unthinkability

edges into irresponsibility and incoherence, whereas thinkability collapses into the normal—which is to say the nontraumatic." What trauma theory must do, says Rubenstein—and this is just as true for ecological trauma—is to continue bridging the gap between the thinkable and the unthinkable, "tacking back and forth between these poles." See Rubenstein, "Afterword," in *Trauma and Transcendence*, ed. Eric Boynton and Peter Capretto (New York: Fordham University Press, 2018), 283–94, 286, 289.

47. Southgate, *The Groaning of Creation*, 16. In *The Groaning of Creation*, Southgate moves from natural theodicy back into environmental ethics when he suggests that human beings might have a role as co-redeemers through, for example, dietary changes or by seeking to reduce the rate of natural extinctions. See Southgate, *The Groaning of Creation*, 116–33. Trauma theology offers a different, and parallel, way forward.

48. Deane-Drummond, *Eco-Theology*, 118.

49. Kenneth Surin, *Theology and the Problem of Evil* (Oxford: Blackwell, 1986); Terrence W. Tilley, *The Evils of Theodicy* (Washington, D.C.: Georgetown University Press, 1991); John Swinton, *Raging with Compassion: Pastoral Responses to the Problem of Evil* (Grand Rapids, MI: Eerdmans, 2007), especially 1–29.

50. Wendy Farley, *Tragic Vision and Divine Compassion: A Contemporary Theodicy* (Louisville, KY: Westminster John Knox Press, 1990), 22.

51. Farley, *Tragic Vision and Divine Compassion*, 23.

52. John Foster, *After Sustainability: Denial, Hope, Retrieval* (Abingdon: Routledge, 2015), 93–95.

53. Panu Pihkala glosses this point as follows: "By using terminology related to tragedy, it is possible to show understanding for the great losses that have already occurred and will inevitably still occur, although humanity can yet have an effect on how much damage climate change will bring in the future." Tragedy is unavoidable, but we do still have some control over the scale of the damage. See Pihkala, "Eco-Anxiety, Tragedy, and Hope: Psychological and Spiritual Dimensions of Climate Change," *Zygon* 53 (2018): 545–69, 554.

54. Foster, *After Sustainability*, 111.

55. Pihkala, "Eco-Anxiety, Tragedy, and Hope," 555.

56. Flora A. Keshgegian, *Time for Hope: Practices for Living in Today's World* (New York: Continuum, 2006), 111.

57. Rambo, *Spirit and Trauma*, 5.

58. Donna M. Orange, *Climate Crisis, Psychoanalysis and Radical Ethics* (Abingdon: Routledge, 2017), xv.

59. Bethany N. Sollereder, "Compassionate Theodicy: A Suggested Truce Between Intellectual and Practical Theodicy," *Modern Theology* 37 (2021): 382–95, 386.

60. Pamela R. McCarroll, "Embodying Theology: Trauma Theory, Climate Change, Pastoral and Practical Theology," *Religions* 13 (2022): 1–14. See also: Ryan LaMothe, "This Changes Everything: The Sixth Extinction and Its Implications for

Pastoral Theology," *Journal of Pastoral Theology* 26 (2016): 178–94, 186; Storm Swain, "Climate Change and Pastoral Theology," in *T & T Clark Handbook of Christian Theology and Climate Change*, ed. Ernst M. Conradie and Hilda P. Koster (London: T & T Clark, 2020), 615–26, 623–24; Bonnie J. Miller-McLemore, "Climate Violence and Earth Justice: A Research Report on Practical Theology's Contributions," *International Journal of Practical Theology* 26 (2022): 329–66, 346–51.

61. McCarroll, "Embodying Theology," 2.
62. McCarroll, "Embodying Theology," 7.
63. McCarroll, "Embodying Theology," 7.
64. McCarroll, "Embodying Theology," 7.
65. See, for example: Francis, *Laudato Si'*, §116, 236; Mark D. Liederbach, "Stewardship: A Biblical Concept?," in *The Oxford Handbook of the Bible and Ecology*, ed. Hilary Marlow and Mark Harris (Oxford: Oxford University Press, 2022), 310–23.
66. McCarroll, "Embodying Theology," 10.
67. Keller, *Political Theology of the Earth*, 93, 95, 108. Catherine Keller also employs the language of trauma in her book *Facing Apocalypse*. She talks here, in the context of a reflection on the Revelation to John, of both "ecological trauma" and "earth trauma." She links "ecological trauma" to the prospect of "de-creation," but the terms themselves do not receive any detailed development. See Keller, *Facing Apocalypse: Climate, Democracy, and Other Last Chances* (Maryknoll, NY: Orbis Books, 2021), 46, 69.
68. Keller, *Political Theology of the Earth*, 93.
69. Keller, *Political Theology of the Earth*, 93.
70. Keller, *Political Theology of the Earth*, 88, quoting from Shelly Rambo, *Resurrecting Wounds: Living in the Afterlife of Trauma* (Waco, TX: Baylor University Press, 2017), 2.
71. Keller, *Political Theology of the Earth*, 93.
72. Rambo, *Resurrecting Wounds*, 4, quoted in Keller, *Political Theology of the Earth*, 93.
73. Danielle Tumminio Hansen, "The Body of God, Sexually Violated: A Trauma-Informed Reading of the Climate Crisis," *Religions* 13 (2022): 1–12, 1.
74. Sallie McFague, *Models of God: Theology for an Ecological, Nuclear Age* (London: SCM Press, 1987), 60.
75. Tumminio Hansen, "The Body of God, Sexually Violated," 8.
76. Tumminio Hansen, "The Body of God, Sexually Violated," 8.
77. For a succinct summary of the issues and a justification of the passibilist position see: Southgate, *The Groaning of Creation*, 56–57.
78. McFague, *Models of God*, 72.
79. Mark I. Wallace, "The Wounded Spirit as the Basis for Hope in an Age of Radical Ecology," in *Christianity and Ecology: Seeking the Well-Being of Earth and Humans*, ed. Dieter T. Hessel and Rosemary Radford Ruether (Cambridge, MA: Harvard University Press, 2000), 51–72, 60.

80. Mark I. Wallace, *When God Was a Bird* (New York: Fordham University Press, 2019), 14–15, referencing Luke 3:21–22.

81. Wallace, "The Wounded Spirit," 60.

82. Wallace, "The Wounded Spirit," 61.

83. In fact, Wallace suggests that McFague's metaphor does not go far enough. McFague writes that "the universe is dependent on God in a way that God is not dependent on the universe." Yet Wallace is not convinced by the dual assertion that the world is God's body, but that God is not dependent on the world. In McFague's scheme, says Wallace, "God is not vulnerable to loss and destruction in the event that God's earth-body is destroyed." See Sallie McFague, *The Body of God: An Ecological Theology* (London: SCM Press, 1993), 149; Mark I. Wallace, *Fragments of the Spirit: Nature, Violence, and the Renewal of Creation* (New York: Continuum, 1996), 140–41.

84. Tumminio Hansen, "The Body of God, Sexually Violated," 5.

85. Tumminio Hansen, "The Body of God, Sexually Violated," 5.

86. Wallace, *When God Was a Bird*, 15.

87. Wallace, *When God Was a Bird*, 166. Wallace also expresses the same idea when he writes elsewhere: "Jesus' body as God's enfleshed presence was inscribed with the marks of human sin; so also the earth body of the Spirit is lacerated by continual assaults upon our planet home. Consider the sad parallels between the crucified Jesus and the cruciform Spirit. The lash marks of human sin cut into the body of the crucified God are now even more graphically displayed across the expanse of the whole planet, as the body of the wounded Spirit bears the incisions of continual abuse." See Wallace, "The Wounded Spirit," 62.

88. However, Eleanor Rae finds the parallel between the suffering Earth and the suffering of Christ problematic: Christ only suffered a few hours, whereas parts of the Earth have suffered for millennia. See Rae, "Response to Mark I. Wallace: Another View of the Spirit's Work," in *Christianity and Ecology: Seeking the Well-Being of Earth and Humans*, ed. Dieter T. Hessel and Rosemary Radford Ruether (Cambridge, MA: Harvard University Press, 2000), 76–77. Furthermore, Christ arguably had some choice in his suffering, whereas many parts of the Earth do not. See Southgate, *The Groaning of Creation*, 50.

89. Mark I. Wallace, "Elegy for a Lost World," in *Post-Traumatic Public Theology*, ed. Stephanie N. Arel and Shelly Rambo (Cham, Switzerland: Palgrave Macmillan, 2016), 152n16.

90. Matthew Eaton, "Conclusion: Ecocide as Deicide: Eschatological Lamentation and the Possibility of Hope," in *Integral Ecology for a More Sustainable World: Dialogues with Laudato Si'*, ed. Dennis O'Hara et al. (Lanham, MD: Lexington Books, 2019), 359.

91. Eaton, "Conclusion," 359. He continues, "it would seem there is less on the line, and less urgency required to act, if the violence we commit now is simply wiped away without its consequences reverberating eschatologically." See Eaton, "Conclusion," 362.

92. Eaton, "Conclusion," 360. See Francis, *Laudato Si'*, §61.
93. Eaton, "Conclusion," 362.
94. Francis, *Laudato Si'*, §33. See also: Eaton, "Conclusion," 362–63.
95. Eaton, "Conclusion," 364. By ecocide and deicide, Eaton does not mean the annihilation of the planet or the annihilation of God, but domination of the planet and the eschatological trauma of God.

3. Ecology in Trauma Theology

1. Serene Jones, *Trauma and Grace: Theology in a Ruptured World*, 2nd Edition (Louisville, KY: Westminster John Knox Press, 2019), xiv.
2. Jones, *Trauma and Grace*, xiv.
3. One explanation for the relative lack of engagement with the ecological realm to date is that trauma theologians have sometimes inherited a residual anthropocentrism from their theological interlocutors. For example, Jones's writing is informed by the thought of John Calvin. She talks poignantly about the trauma of terrorist attacks, domestic abuse, racial discrimination, and pregnancy loss, but ecological issues are notable for their absence. When Jones does invoke the rest of creation, she follows Calvin in describing it as "the theater of God's glory," relegating nonhuman entities to the role of providing a stage for divine and human dramas. Calvin's understanding of creation is undoubtedly nuanced, yet he often indicates that Christ only became incarnate to redeem fallen humanity (not the rest of creation), and that any beauty in creation is a revelation of God rather than an indication of intrinsic value. Calvin's limited engagement with the nonhuman creation on its own terms may therefore subtly shape the focus of Jones's trauma theology. See Jones, *Trauma and Grace*, 43–67, 104; Stephen Edmondson, *Calvin's Christology* (Cambridge: Cambridge University Press, 2004), 53, 145–46; Susan E. Schreiner, *The Theater of His Glory: Nature and the Natural Order in the Thought of John Calvin*, Studies in Historical Theology (Durham, NC: Labyrinth Press, 1991); Serene Jones, *Calvin and the Rhetoric of Piety*, 1st Edition, Columbia Series in Reformed Theology (Louisville, KY: Westminster John Knox Press, 1995).
4. Deanna A. Thompson, *Glimpsing Resurrection: Cancer, Trauma, and Ministry* (Louisville, KY: Westminster John Knox Press, 2018), 5, 43–46. Thompson writes, "while conventional understandings of traumatic events focus on extraordinary occurrences in the past that have a beginning, middle, and end, trauma associated with illness typically does not arise from a single event but rather from recurring events extending from diagnosis through treatment and beyond, possibly throughout the rest of a person's life." See Thompson, *Glimpsing Resurrection*, 6.
5. Thompson, *Glimpsing Resurrection*, 7, 18–19, 48–49. As Thompson puts it, "not all undoing is caused by immoral human acts. Some undoings work from the inside out; the threat emerges internally and undoes not just the body but the psyche and the spirit." See Thompson, *Glimpsing Resurrection*, 41.

6. Kai Erikson, *A New Species of Trouble: Explorations in Disaster, Trauma, and Community* (New York: W. W. Norton, 1994), 227.

7. Kai Erikson, *Everything in Its Path: Destruction of Community in the Buffalo Creek Flood* (New York: Simon & Schuster, 1976), 154.

8. Kai Erikson, "Notes on Trauma and Community," in *Trauma: Explorations in Memory*, ed. Cathy Caruth (Baltimore, MD: Johns Hopkins University Press, 1995), 183–99, 185. It is noteworthy that Jeffrey C. Alexander has proposed a slightly different theory of collective trauma. For Alexander, the trauma resides, not in the pain experienced by the community, but in the way that pain is represented and remembered—what he calls the "trauma process." For Alexander, certain elite "carrier groups" such as journalists, politicians, intellectuals, and religious leaders, who are responsible for meaning making in the public sphere, have control over whether and how a given event is narrated and constructed as a collective (or cultural) trauma. See Alexander, "Toward a Theory of Cultural Trauma," in *Cultural Trauma and Collective Identity*, ed. Jeffrey C. Alexander et al. (Berkeley: University of California Press, 2004), 1–30, 1, 11, 26.

9. Gay Bradshaw, *Elephants on the Edge: What Animals Teach Us About Humanity* (New Haven, CT: Yale University Press, 2009), 78, 81–86.

10. Bradshaw, *Elephants on the Edge*, 108.

11. Stacy M. Lopresti-Goodman et al., "Psychological Distress in Chimpanzees Rescued From Laboratories," *Journal of Trauma and Dissociation* 16 (2015): 349–66, 349.

12. Jessica Deslauriers et al., "Current Status of Animal Models of Posttraumatic Stress Disorder: Behavioral and Biological Phenotypes, and Future Challenges in Improving Translation," *Biological Psychiatry* 83 (2018): 895–907.

13. Stef Craps, "Climate Trauma," in *The Routledge Companion to Literature and Trauma*, ed. Colin Davis and Hanna Meretoja (London: Routledge, 2020), 275–84, 282. In a similar vein, Michael Rothberg proposes that trauma studies must "add a whole new series of destinations to its agenda." He insists on "the necessity . . . of broadening and differentiating our understanding of what trauma is" and specifically mentions how "ecological devastation can be traumatic." See Rothberg, "Preface: Beyond Tancred and Clorinda—Trauma Studies for Implicated Subjects," in *The Future of Trauma Theory: Contemporary Literary and Cultural Criticism*, ed. Gert Buelens et al. (Abingdon: Routledge, 2013), xi–xviii, xii, xvii.

14. Thompson, *Glimpsing Resurrection*, 6–7, 43–46.

15. Jones, *Trauma and Grace*, xii.

16. Jones, *Trauma and Grace*, xiii.

17. Katie Cross and Karen O'Donnell, "Introduction," in *Bearing Witness: Intersectional Perspectives on Trauma Theology*, ed. Karen O'Donnell and Katie Cross (London: SCM Press, 2022), 1–9, 2, 8.

18. Megan Warner et al., eds., *Tragedies and Christian Congregations: The Practical Theology of Trauma* (Abingdon: Routledge, 2019).

19. Shelly Rambo, *Spirit and Trauma: A Theology of Remaining* (Louisville, KY: Westminster John Knox Press, 2010).

20. Shelly Rambo, *Resurrecting Wounds: Living in the Afterlife of Trauma* (Waco, TX: Baylor University Press, 2017), 114.

21. Rambo, *Resurrecting Wounds*, 111, 134.

22. Rambo, *Spirit and Trauma*, 4. Likewise, she observes how, in the history of trauma studies, "trauma moved off the psychoanalytic couch . . . and beyond strictly clinical fields to fields like literature, history, politics, and religion." See Rambo, *Spirit and Trauma*, 26.

23. Rambo, *Resurrecting Wounds*, 4.

24. Roger Luckhurst, *The Trauma Question* (London: Routledge, 2008), 4; Shelly Rambo, "Introduction," in *Post-Traumatic Public Theology*, ed. Stephanie N. Arel and Shelly Rambo (Cham, Switzerland: Palgrave Macmillan, 2016), 1–21, 2; Cathy Caruth, "Trauma and Experience: Introduction," in *Trauma: Explorations in Memory*, ed. Cathy Caruth (Baltimore, MD: Johns Hopkins University Press, 1995), 3–12, 4.

25. Neil J. Smesler, "Psychological Trauma and Cultural Trauma," in *Cultural Trauma and Collective Identity*, ed. Jeffrey C. Alexander et al. (Berkeley: University of California Press, 2004), 31–59, 58.

26. Wulf Kansteiner, "Genealogy of a Category Mistake: A Critical Intellectual History of the Cultural Trauma Metaphor," *Rethinking History* 8 (2004): 193–221, 213.

27. Kansteiner, "Genealogy of a Category Mistake," 214.

28. Rambo specifically discusses the relevance of Rothberg's work in the context of trauma theology. See Rambo, *Resurrecting Wounds*, 95.

29. Michael Rothberg, *Multidirectional Memory: Remembering the Holocaust in the Age of Decolonization*, Cultural Memory in the Present (Stanford, CA: Stanford University Press, 2009), 9.

30. Luckhurst, *The Trauma Question*, 14.

31. Cross and O'Donnell, "Introduction," 2.

32. Sandro Galea et al., "Exposure to Hurricane-Related Stressors and Mental Illness After Hurricane Katrina," *Archives of General Psychiatry* 64 (2007): 1427–34, 1427. See also: Susan Clayton et al., *Mental Health and Our Changing Climate: Impacts, Implications, and Guidance* (Washington, D.C.: American Psychological Association and ecoAmerica, 2017), 22.

33. Kevin U. Stephens et al., "Excess Mortality in the Aftermath of Hurricane Katrina: A Preliminary Report," *Disaster Medicine and Public Health Preparedness* 1 (2007): 15–20.

34. Ron Eyerman, *Is This America?: Katrina as Cultural Trauma* (Austin: University of Texas Press, 2015), 6–8.

35. Rambo, *Spirit and Trauma*, 2.

36. Rambo, *Spirit and Trauma*, 27. As Randy Fertel expresses it, "one sure way to provoke a fistfight in New Orleans these days is to speak of Hurricane Katrina as 'a natural disaster.'" Rather, the catastrophe included "the failure of the federal levees" and "years of neglect from the powers that be," as well as "corporate fecklessness and incompetence." Fertel, "Hearing the Bugle's Call: Hurricane Katrina, the BP Oil

Spill, and the Effects of Trauma," in *Environmental Disasters and Collective Trauma*, ed. Nancy Cater and Stephen Foster (New Orleans: Spring Journal, 2012), 91–115, 94.

37. Shelly Rambo, "Saturday in New Orleans: Rethinking the Holy Spirit in the Aftermath of Trauma," *Review and Expositor* 105 (2008): 229–44, 230.

38. Fertel, "Hearing the Bugle's Call," 94.

39. Rambo, *Spirit and Trauma*, 9–10.

40. Rambo, *Spirit and Trauma*, 2.

41. Erikson, "Notes on Trauma and Community," 183.

42. Lauren Woolbright, "Wounded Planet, Wounded People: The Possibility of Ecological Trauma," unpublished master's thesis (Clemson University, 2011), ii, accessed 28 October 2018, https://www.proquest.com/docview/881256145. Donna Orange goes one step further, highlighting the ethical import of recognizing this double trauma. She writes, "the traumas of others—especially at this moment of climate emergency—must traumatize us (traumatism), if we and they are to be transformed by their transcendence over us." In other words, we are only likely to be spurred into acting on climate change when we recognize that the ecological trauma of climate change both impacts and implicates ourselves. See Orange, "Traumatized by Transcendence," in *Trauma and Transcendence*, ed. Eric Boynton and Peter Capretto (New York: Fordham University Press, 2018), 70–82, 79.

43. Woolbright, "Wounded Planet, Wounded People," 25.

44. As in Chapter 2, I am relying on the work of others who have explicitly demonstrated the collapse of the nature/culture binary. See Philippe Descola, *Beyond Nature and Culture*, trans. Janet Lloyd (Chicago: University of Chicago Press, 2013), 57–88; Bruno Latour, *We Have Never Been Modern*, trans. Catherine Porter (Cambridge, MA: Harvard University Press, 1993), 2, 7.

45. Interestingly, Erikson undermines some of his own sensitivity to the ecological dimensions of trauma by falling foul of the nature/culture dichotomy. He writes, "natural disasters are experienced as acts of God or whims of nature. They occur to us. They visit us, as if from afar. Technological disasters, on the other hand, being of human manufacture, are at least in principle preventable." Erikson trades here on a false distinction between "natural" and "technological" disasters. He even goes so far as to say that "one of the crucial tasks of culture . . . is to help people camouflage the actual risks of the world around them—to help them edit reality in such a way that it seems manageable." Instead of acknowledging human culture as emerging from, and relying on, what is natural, Erikson pits culture against nature. If we continue to conceive of nature as the enemy to be vanquished, or concealed, by culture, then it will be hard to become sympathetic to the trauma in the scene itself. See Erikson, "Notes on Trauma and Community," 191, 194.

46. Rambo, *Spirit and Trauma*, xiii.

47. Rambo, "Introduction," 5. Rambo, unlike Jones, already seems comfortable with the anthropomorphic language of a "vulnerable" Earth and a "wounded" environment.

48. Jones, *Trauma and Grace*, 15.

49. Jennifer Baldwin, *Trauma-Sensitive Theology: Thinking Theologically in the Era of Trauma* (Eugene, OR: Wipf & Stock, 2018), 44.

50. Karen O'Donnell, *Broken Bodies: The Eucharist, Mary, and the Body in Trauma Theology* (London: SCM Press, 2018), 79.

51. Karen O'Donnell, "Eucharist and Trauma: Healing in the B/body," in *Tragedies and Christian Congregations: The Practical Theology of Trauma*, ed. Megan Warner et al. (Abingdon: Routledge, 2019), 182–93, 185.

52. O'Donnell, *Broken Bodies*, 79.

53. Elizabeth Boase and Christopher G. Frechette, "Defining 'Trauma' as a Useful Lens for Biblical Interpretation," in *Bible Through the Lens of Trauma*, ed. Elizabeth Boase and Christopher G. Frechette (Atlanta: SBL Press, 2016), 1–24, 11.

54. Rebecca L. Copeland, "'Their Leaves Shall Be for Healing': Ecological Trauma and Recovery in Ezekiel 47:1–12," *Biblical Theology Bulletin* 49 (2019): 214–22, 214.

55. Copeland, "Their Leaves Shall Be for Healing," 215.

56. Copeland, "Their Leaves Shall Be for Healing," 215.

57. Copeland, "Their Leaves Shall Be for Healing," 219. See also the discussion of Hosea in Chapter 1.

58. Megan Warner, "Teach to Your Daughters a Dirge: Revisiting the Practice of Lament in the Light of Trauma Theory," in *Tragedies and Christian Congregations: The Practical Theology of Trauma*, ed. Meg Warner et al. (Abingdon: Routledge, 2019), 167–81, 170. See also: Thompson, *Glimpsing Resurrection*, 75.

59. Brent A. Strawn, "Trauma, Psalmic Discourse, and Authentic Happiness," in *Bible Through the Lens of Trauma*, ed. Elizabeth Boase and Christopher G. Frechette (Atlanta: SBL Press, 2016), 143–60, 152.

60. Psalm 77:4.

61. Psalm 38:7–8.

62. Warner, "Teach to Your Daughters a Dirge," 172.

63. Warner, "Teach to Your Daughters a Dirge," 175.

64. Thompson, *Glimpsing Resurrection*, 84.

65. David Rensberger, "Ecological Use of the Psalms," in *The Oxford Handbook of the Psalms*, ed. William P. Brown, Oxford Handbooks (Oxford: Oxford University Press, 2014), 608–19, 608–9.

66. For instance, in Ezekiel "the land was appalled"; in Jeremiah "the earth shall mourn"; and in Hosea the mourning of the animals and the Earth sets an example for humanity to follow. See Ezekiel 19:7; Jeremiah 4:28; Hosea 4:3. See also: Keith Carley, "Ezekiel's Formula of Desolation: Harsh Justice for the Land/Earth," in *The Earth Story in the Psalms and the Prophets*, ed. Norman C. Habel (Sheffield: Sheffield Academic Press, 2001), 143–57, 154; Shirley Wurst, "Retrieving Earth's Voice in Jeremiah: An Annotated Voicing of Jeremiah 4," in *The Earth Story in the Psalms and the Prophets*, ed. Norman C. Habel (Sheffield: Sheffield Academic Press, 2001), 172–84, 183; Laurie J. Braaten, "Earth Community in Hosea 2," in *The Earth Story in the Psalms and the Prophets*, ed. Norman C. Habel (Sheffield: Sheffield Academic

Press, 2001), 185–203, 203; Melissa Tubbs Loya, "Therefore the Earth Mourns: The Grievance of Earth in Hosea 4:1–3," in *Exploring Ecological Hermeneutics*, ed. Norman C. Habel and Peter Trudinger (Atlanta: Society of Biblical Literature, 2008), 53–62, 56.

67. Isaiah 24:19.

68. Jones, *Trauma and Grace*, xi.

69. Hilary Jerome Scarsella, "Trauma and Theology," in *Trauma and Transcendence*, ed. Eric Boynton and Peter Capretto (New York: Fordham University Press, 2018), 256–82, 256.

70. Matthew 27:46 and Mark 15:34, quoting Psalm 22:1.

71. See, for example, James H. Cone's *The Cross and the Lynching Tree*, where Christ's crucifixion bears witness to the terror and trauma of Black lynchings. In Cone's words, "God transformed lynched black bodies into the recrucified body of Christ," thereby bringing Christ's crucifixion trauma into the heart of Black experience. See Cone, *The Cross and the Lynching Tree* (Maryknoll, NY: Orbis Books, 2011), 158.

72. Matthew 27:45, 51. See also: Mark 15:33; Luke 23:44.

73. Norman C. Habel with the Earth Bible Team, "Where Is the Voice of Earth in Wisdom Literature?," in *The Earth Story in Wisdom Traditions*, ed. Norman C. Habel and Shirley Wurst (Sheffield: Sheffield Academic Press, 2001), 23–34, 33.

74. Elaine Mary Wainwright, *Habitat, Human, and Holy: An Eco-Rhetorical Reading of the Gospel of Matthew*, Earth Bible Commentary Series (Sheffield: Sheffield Phoenix Press, 2016), 212.

75. Alan Cadwallader, "'And the Earth Shook'—Mortality and Ecological Diversity: Interpreting Jesus' Death in Matthew's Gospel," in *Biodiversity and Ecology as Interdisciplinary Challenge*, ed. Denis Edwards and Mark William Worthing (Adelaide: ATF Press, 2004), 45–54, 53.

76. Wainwright, *Habitat, Human, and Holy*, 211.

77. Jones, *Trauma and Grace*, xiv.

4. The Rupture of Communication: Christ's Witness to a Wounded World

1. Donna J. Haraway, *Staying with the Trouble: Making Kin in the Chthulucene* (Durham, NC: Duke University Press, 2016), 1.

2. Haraway, a non-theologian, would likely be content to dismiss eschatology altogether. Many trauma theologians also tend to leave eschatology out of their discussions. For example, Serene Jones warns against the use of eschatology as "a facile palliative for those who mourn." See Jones, *Trauma and Grace: Theology in a Ruptured World*, 2nd Edition (Louisville, KY: Westminster John Knox Press, 2019), 145. Nevertheless, a more nuanced eschatology that commits to staying with the trouble may yet be possible in cases of trauma.

3. Haraway, *Staying with the Trouble*, 10. Heather Eaton expresses a cognate sentiment in her exposition of the concept of "planetary solidarity." She writes, "a

dedication to reality as encountered rather than a fantasy of another life elsewhere might be a better place from which to develop planetary solidarity." See Eaton, "An Earth-Centric Theological Framing for Planetary Solidarity," in *Planetary Solidarity: Global Women's Voices on Christian Doctrine and Climate Justice*, ed. Grace Ji-Sun Kim and Hilda P. Koster (Minneapolis: Fortress Press), 19–44, 44.

4. Haraway, *Staying with the Trouble*, 1, 10, 137. The one place where I think Haraway's analysis goes a step too far is when she says that we should not "succumb" to hope, since it is not a "sensible attitude." See Haraway, *Staying with the Trouble*, 4. The problem is that Haraway seems to be conflating hope with optimism. In dismissing problematically triumphant futures, Haraway risks jettisoning the "partial recuperation" and "germs of partial healing" that she herself wants to retain. If Haraway is to truly reject fatalism, then the practice of staying with the trouble must leave a small opening for a chastened form of hope. See Haraway, *Staying with the Trouble*, 137. I also take up this discussion of hope in greater detail in the Conclusion.

5. Shelly Rambo, *Spirit and Trauma: A Theology of Remaining* (Louisville, KY: Westminster John Knox Press, 2010), 18; Karen O'Donnell, *Broken Bodies: The Eucharist, Mary and the Body in Trauma Theology* (London: SCM Press, 2018), 6–7.

6. Rambo, *Spirit and Trauma*, 18.

7. Rambo, *Spirit and Trauma*, 21; O'Donnell, *Broken Bodies*, 7.

8. Dori Laub, "Truth and Testimony: The Process and the Struggle," in *Trauma: Explorations in Memory*, ed. Cathy Caruth (Baltimore, MD: Johns Hopkins University Press, 1995), 61–75, 65.

9. I am grateful to Kimberly Carfore for pointing out this resonance between the idea of bearing witness and Haraway's notion of staying with the trouble when I presented some of these ideas at the American Academy of Religion annual meeting in 2020. Catherine Keller is also attuned to the way that Rambo's remaining witness "obliquely resembles" Haraway's professed commitment. See Keller, *Political Theology of the Earth: Our Planetary Emergency and the Struggle for a New Public* (New York: Columbia University Press, 2018), 88.

10. Rambo, *Spirit and Trauma*, 38.

11. Rambo, *Spirit and Trauma*, 22.

12. Rambo, *Spirit and Trauma*, 39.

13. See Elizabeth Castelli, *Imitating Paul: A Discourse of Power*, Literary Currents in Biblical Interpretation (Louisville, KY: Westminster John Knox Press, 1991), 21.

14. See Elizabeth Castelli, *Martyrdom and Memory: Early Christian Culture Making*, Gender, Theory, and Religion (New York: Columbia University Press, 2004), 203; Rambo, *Spirit and Trauma*, 16.

15. See Mary Daly, *Beyond God the Father: Towards a Philosophy of Women's Liberation* (Wellingborough: The Women's Press, 1986), 77; Elisabeth Schüssler Fiorenza, "Proclaimed by Women: The Execution of Jesus and the Theology of the Cross," in *Jesus: Miriam's Child, Sophia's Prophet: Critical Issues in Feminist*

Christology (London: Bloomsbury T & T Clark, 2015), 105–39, 111; Delores S. Williams, *Sisters in the Wilderness: The Challenge of Womanist God-Talk* (Maryknoll, NY: Orbis Books, 1993), 143.

16. See Chapter 2 for further discussion of the unassimilable nature of ecological trauma.

17. Rambo, *Spirit and Trauma*, 2, 13. Rambo builds here on Rebecca S. Chopp's work on the "poetics of testimony." This form of witness, in contrast to the traditional models of proclamation and imitation, does not have to *conform* to preexisting theological norms but is rather supposed to *inform* the task of theological discourse. Theology is not an external arbiter of witnessing, but a respondent to the moral summons of testimony. See Chopp, "Theology and the Poetics of Testimony," in *Converging on Culture: Theologians in Dialogue with Cultural Analysis and Criticism*, ed. Delwin Brown et al. (Oxford: Oxford University Press, 2001), 56–70; Rambo, *Spirit and Trauma*, 165.

18. Rambo, *Spirit and Trauma*, 26.

19. Rambo, *Spirit and Trauma*, 104, based on the OED definition of "remaining."

20. Rambo, *Spirit and Trauma*, 102–4. See also: John 15:4–10.

21. Kelly Oliver, *Witnessing: Beyond Recognition* (Minneapolis: University of Minnesota Press, 2001), 16.

22. Oliver, *Witnessing: Beyond Recognition*, 85, 99.

23. Blanche Verlie, *Learning to Live with Climate Change: From Anxiety to Transformation*, Routledge Focus on Environment and Sustainability (London: Routledge, 2022), 12.

24. Verlie, *Learning to Live with Climate Change*, 68.

25. Verlie, *Learning to Live with Climate Change*, 68.

26. Verlie, *Learning to Live with Climate Change*, 72.

27. Verlie, *Learning to Live with Climate Change*, 119.

28. Rambo, *Spirit and Trauma*, 111–41.

29. Rambo, *Spirit and Trauma*, 116.

30. For example, says Rambo, "they interpret the significance of redemption through a christological model of imitation and self-sacrifice that cannot, I claim, bear the tensions of death and life as they exist on Holy Saturday. . . . They increasingly link Holy Saturday to a christocentric logic that secures rather than testifies. The subtext of witness in their thought . . . is elided in favour of an imitative model of martyrdom that replicates while expanding a certain logic of the passion." See Rambo, *Spirit and Trauma*, 48, 62, 68.

31. Rambo, *Spirit and Trauma*, 70.

32. Rambo, *Spirit and Trauma*, 69.

33. Balthasar even writes in *The Moment of Christian Witness* that "martyrdom means bearing witness." See Hans Urs von Balthasar, *The Moment of Christian Witness*, trans. Richard Beckley, Communio Books (San Francisco: Ignatius Press, 1994), 142, quoted in Rambo, *Spirit and Trauma*, 70.

34. Rambo, *Spirit and Trauma*, 112.

35. Rambo, *Spirit and Trauma*, 71.

36. Preston Hill, "Does God Need a Body to Keep the Score of Trauma?," *Theological Puzzles* (2021), accessed 9 January 2021, https://www.theo-puzzles.ac.uk/2021/04/20/phill/.

37. Celia Deane-Drummond detects the same problem in Balthasar's original treatment of Christ's descent into hell. She writes, "the difficulty, of course, is how far Balthasar's speculations about Christ confronting absolute sin in Hell represent an unfortunate type of dis-incarnation, a removal from the Word made *flesh*." If Christ is dead, and has descended to hell, has the *Logos* escaped from the world of material flesh at precisely the moment when we most require an embodied form of divine solidarity and witness? Rambo's shift to the middle Spirit for the duration of Holy Saturday does little to allay this concern about disembodiment. See Deane-Drummond, "Who on Earth Is Jesus Christ? Plumbing the Depths of Deep Incarnation," in *Christian Faith and the Earth: Current Paths and Emerging Horizons in Ecotheology*, ed. Ernst M. Conradie et al. (New York: T & T Clark, 2014), 31–50, 44.

38. John 19:30; Rambo, *Spirit and Trauma*, 118.

39. Rambo, *Spirit and Trauma*, 119.

40. Hill, "Does God Need a Body to Keep the Score of Trauma?," para. 14.

41. Traditionally, Christ's body is understood to continue in the eucharist and in the church. Both of these bodies can also serve as witnesses to the trauma at the heart of Christianity, and I will explore them briefly in the Conclusion. However, these "bodies of Christ" are best able to serve as witnesses if Christ himself, and not the Spirit, is understood as the primary witness to trauma. The subsequent displacements of Christ's body, and the way in which this might destabilize the identity of the body of Christ, are explored in an essay by Graham Ward. See Ward, "The Displaced Body of Jesus Christ," in *Radical Orthodoxy: A New Theology*, ed. John Milbank et al. (London: Routledge, 1998), 163–81.

42. Rambo, *Spirit and Trauma*, 122. Rambo draws here on Keller's pneumatological materiality. See Catherine Keller, *Face of the Deep: A Theology of Becoming* (Abingdon: Routledge, 2003), 233. Mark I. Wallace also offers several arguments for the physicality (and animality) of the Spirit. See, for example: Wallace, *When God Was a Bird* (New York: Fordham University Press, 2019).

43. As Wallace admits, "discourse about spirit remains saddled with ethereal and pejorative connotations, conjuring images of ghosts, phantoms, and other incorporeal forces; of vaporous clouds and gaseous substances; of whatever is airy, immaterial, invisible, nonsubstantial, bloodless, bodiless, passionless and unearthly." These connotations make it very hard to imagine the Spirit as an enfleshed witness. See Mark I. Wallace, "The Wounded Spirit as the Basis for Hope in an Age of Radical Ecology," in *Christianity and Ecology: Seeking the Well-Being of Earth and Humans*, ed. Dieter T. Hessel and Rosemary Radford Ruether (Cambridge, MA: Harvard University Press, 2000), 51–72, 54–55.

44. This is not to say that the Spirit cannot be part of Christ's witness. Indeed,

this must be true to the extent that all persons of the Trinity are present in all its operations. The tension between pneumatological and Christological witnessing could be resolved within a fully trinitarian theology.

45. Hill, "Does God Need a Body to Keep the Score of Trauma?," para. 27.

46. For example, Richard Bauckham, *Jesus and the Eyewitnesses: The Gospels as Eyewitness Testimony*, 2nd Edition (Grand Rapids, MI: Eerdmans, 2017).

47. When Christ explains his purpose to Pilate in John's gospel, he says, "for this cause came I into the world, that I should bear witness unto the truth." But "truth" here is ambiguous: it could refer to the truth of God or the truth of the suffering world—or both, since the two are not mutually exclusive. See John 18:37, KJV.

48. Jon Sobrino, *Christ the Liberator: A View from the Victims*, trans. Paul Burns (Maryknoll, NY: Orbis Books, 2001), 138, 148.

49. Jon Sobrino, *Jesus the Liberator: A Historical-Theological Reading of Jesus of Nazareth*, trans. Paul Burns and Francis McDonagh (Maryknoll, NY: Orbis Books, 1993), 244.

50. Jakub Urbaniak pinpoints the problem precisely. He writes, "even christologizing centred upon the category of 'flesh' is at risk of remaining purely visionary unless it is done by and/or with those in whose bodies Jesus is being crucified." See Urbaniak, "Extending and Locating Jesus's Body: Toward a Christology of Radical Embodiment," *Theological Studies* 80 (2019): 774–97, 774.

51. Elizabeth A. Johnson, "Jesus and the Cosmos: Soundings in Deep Christology," in *Incarnation: On the Scope and Depth of Christology*, ed. Niels Henrik Gregersen (Minneapolis: Fortress Press, 2015), 133–56, 140.

52. Matthew 6:26, 28.

53. See, for example: Niels Henrik Gregersen, "Deep Incarnation: Why Evolutionary Continuity Matters in Christology," *Toronto Journal of Theology* 26 (2010): 173–88, 182. However, it is worth noting that, despite the ecological appeal of this text, birds and lilies still rank below human beings in Jesus's hierarchy of concern. See Matthew 6:30.

54. Sallie McFague, *Super, Natural Christians: How We Should Love Nature* (Minneapolis: Fortress Press, 1997), 27.

55. Matthew 6:34.

56. Matthew 10:29. The same parable can be found in Luke 12:6. Again, Jesus still places humans much higher in the pecking order, being "of more value than many sparrows." See Matthew 10:31.

57. Denis Edwards, "Every Sparrow That Falls to the Ground: The Cost of Evolution and the Christ-Event," *Ecotheology* 11 (2006): 103–23, 104.

58. Matthew 10:26.

59. Margaret Daly-Denton, *John, an Earth Bible Commentary: Supposing Him to Be the Gardener*, Earth Bible Commentary Series (London: Bloomsbury Academic, 2017).

60. Gregersen, "Deep Incarnation: Why Evolutionary Continuity Matters in Christology," 182; Niels Henrik Gregersen, "The Idea of Deep Incarnation: Biblical

and Patristic Resources," in *To Discern Creation in a Scattering World*, ed. Frederiek Depoortere and Jacques Haers (Leuven: Uitgeverij Peeters, 2013), 319–41, 329.

61. Elizabeth A. Johnson, *Ask the Beasts: Darwin and the God of Love* (London: Bloomsbury, 2014), 200.

62. Denis Edwards, *Ecology at the Heart of Faith* (Maryknoll, NY: Orbis Books, 2006), 50–51.

63. C. H. Dodd, *The Parables of the Kingdom*, Revised Edition (London: Nisbet, 1961), 20–21.

64. Quoted in Jones, *Trauma and Grace*, 76. In Shelly Rambo's words, we can think of the crucifixion as "not only the suffering of one body but also of a body that takes in histories of suffering and bears the marks of these histories." See Rambo, *Resurrecting Wounds: Living in the Afterlife of Trauma* (Waco, TX: Baylor University Press, 2017), 150.

65. Jones, *Trauma and Grace*, 82. Yet the language of mirroring also harbors a potential problem: how do we avoid Christ becoming a product of our own projections? If Christ is merely a mirror, then what is to prevent us from creating Christ in our own image? This is where the idea of witnessing is preferable to that of mirroring: witnessing can be understood as an active process, whereas mirroring is passive and runs the risk of becoming trapped within the orbit of our own desires.

66. Jones, *Trauma and Grace*, 123.

67. This is very much in keeping with Sobrino's "silent witness," mentioned above. See Sobrino, *Jesus the Liberator*, 244.

68. Jones, *Trauma and Grace*, 85–97.

69. One very tangible and practical example of Christ's cruciform witness to ecological suffering is provided by a project from the St. Columban's Mission Society entitled "Stations of the Forests." In a short film and accompanying booklet, the fourteen stations of Christ's passion are recast as akin to the damage wrought by contemporary deforestation. See Timothy A. Middleton, "Christic Witnessing: A Practical Response to Ecological Trauma," *Practical Theology* 15 (2022): 420–31.

70. Johnson, *Ask the Beasts*, 205; Elizabeth A. Johnson, *Creation and the Cross: The Mercy of God for a Planet in Peril* (Maryknoll, NY: Orbis Books, 2018), 188.

71. Rephrased from Jones, *Trauma and Grace*, 123.

72. Elaine Mary Wainwright, *Habitat, Human, and Holy: An Eco-Rhetorical Reading of the Gospel of Matthew*, Earth Bible Commentary Series (Sheffield: Sheffield Phoenix Press, 2016), 211.

73. For Christians, the cruciform Christic witness is clearly central. But we need not presume that Christ is a unique and irreducible mediator of humanity's relationship to the suffering Earth. Traumatized though it is, the nonhuman realm may yet bear witness to its own suffering, and instigate its own form of revelation. See Trevor Bechtel et al., "Introduction," in *Encountering Earth: Thinking Theologically With a More-Than-Human World*, ed. Trevor Bechtel et al. (Eugene, OR: Cascade Books, 2018), 1–13, 4.

74. Arlen Gray, privately published meditation, quoted in Johnson, *Ask the Beasts*, 206.

75. Christopher Southgate provides a potted history of the concept of cruciform creation, identifying its first use in a 1927 work by Charles Raven. However, Southgate suggests that Rolston developed his idea of cruciformity independently of any previous writings on the subject and it is Rolston's name that is most frequently associated with the idea today. See Southgate, *The Groaning of Creation: God, Evolution, and the Problem of Evil* (Louisville, KY: Westminster John Knox Press, 2008), 152–53n37. That said, the idea that the Greek letter X (chi)—a possible reference to the cross of Christ—was imprinted on the universe can be traced back to Irenaeus, Justin Martyr, and Plato. See Denis Minns, *Irenaeus: An Introduction* (London: T & T Clark, 2010), 109.

76. Holmes Rolston III, *Science & Religion: A Critical Survey* (Philadelphia: Templeton Foundation Press, 2006), xl. See also: Holmes Rolston III, *Genes, Genesis and God: Values and Their Origins in Natural and Human History* (Cambridge: Cambridge University Press, 1999), 303–7.

77. Rolston, *Science & Religion*, 291.

78. Rolston, *Science & Religion*, 146.

79. Rolston, *Science & Religion*, 143; Rolston, *Genes, Genesis and God*, 306.

80. Celia Deane-Drummond, *Christ and Evolution: Wonder and Wisdom* (London: SCM Press, 2009), 172.

81. Rolston, *Science & Religion*, 145.

82. Southgate, *The Groaning of Creation*, 44. This is in stark contrast to the parable of the sparrows. See Edwards, "Every Sparrow That Falls to the Ground."

83. For example, "the element of suffering and tragedy is always there," says Rolston, "but it is muted and transmuted in the systemic whole." See Rolston, *Science & Religion*, 137.

84. Edwards, "Every Sparrow That Falls to the Ground," 108.

85. Ruth Page, *God and the Web of Creation* (London: SCM Press, 1996), 82. Page employs the Heideggerian term *Mitsein*, being with, as part of her argument here. See Page, *God and the Web of Creation*, 42.

86. Ruth Page, *Ambiguity and the Presence of God* (London: SCM Press, 1985), 209. What companionship evokes, says Page, is the sense of "being alongside in unoppressive solidarity." See Page, *Ambiguity and the Presence of God*, 207.

87. Page, *God and the Web of Creation*, 155. Page also refers to this idea as "teleology now!" See Page, *God and the Web of Creation*, 63.

88. Page, *Ambiguity and the Presence of God*, 188.

89. Page, *God and the Web of Creation*, 169.

90. Matthew 1:23. Page resists reference to Christ in her work on companionship because, she says, it would involve "too many extra areas of argument." And she worries that "references to Jesus Christ may be clouded by undisclosed assumptions." See Page, *Ambiguity and the Presence of God*, 7. However, she does develop her

Christology in a later work: Ruth Page, *The Incarnation of Freedom and Love* (London: SCM Press, 1991).

91. Arthur Peacocke, "Biological Evolution: A Positive Theological Appraisal," in *Evolutionary and Molecular Biology: Scientific Perspectives on Divine Action*, ed. Robert John Russell et al. (Vatican City: Vatican Observatory Publications, 1998), 357–76, 371–72; Jay B. McDaniel, *Of God and Pelicans: A Theology of Reverence for Life* (Louisville, KY: Westminster John Knox Press, 1989), 29–31.

92. Johnson, *Ask the Beasts*, 191.

93. Johnson, *Ask the Beasts*, 206. Johnson adduces various scriptural warrants for a theology of co-suffering, for example, when God says: "I know their sufferings"; "I drench you with my tears"; "my heart moans for Moab like a flute." See Exodus 3:7; Isaiah 16:9; Jeremiah 48:36; Johnson, *Ask the Beasts*, 192, 203. As Johnson states, "the biblical God has always had compassionate knowledge of creation's suffering." See Johnson, *Ask the Beasts*, 203.

94. Southgate, *The Groaning of Creation*, 52.

95. Johnson, *Creation and Cross*, 189; Christopher Southgate, "Does God's Care Make Any Difference? Theological Reflection on the Suffering of God's Creatures," in *Christian Faith and the Earth: Current Paths and Emerging Horizons in Ecotheology*, ed. Ernst M. Conradie et al. (New York: Bloomsbury T & T Clark, 2014), 97–114, 111. For sentient creatures there is also the possibility that divine co-suffering could reduce meta-level suffering, that is, the additional existential torment of feeling that nobody seems to care about the suffering. Ecologically, where levels of sentience are a topic of debate, co-suffering can still "bring vital divine presence even closer to every dead bird." See Johnson, *Creation and Cross*, 189.

96. Niels Henrik Gregersen, "The Cross of Christ in an Evolutionary World," *Dialog: A Journal of Theology* 40 (2001): 192–207, 205.

97. Denis Edwards, *Deep Incarnation: God's Redemptive Suffering with Creatures* (Maryknoll, NY: Orbis Books, 2019), 123. Note how "God's Redemptive Suffering with Creatures" is in the very subtitle of Edwards's book.

98. Johnson, *Ask the Beasts*, 207; Southgate, "Does God's Care Make Any Difference?," 101–4. The other components of Southgate's compound evolutionary theodicy are: a recognition that there are some logical constraints on what and how God can create (an "only way" argument); eschatological recompense for the suffering of creatures; and a call on humanity to be co-redeemers. See Southgate, *The Groaning of Creation*, 15–17.

99. To put this point in another idiom, can divine solidarity liberate? Justin Ashworth proposes that it does. Ashworth argues that, according to liberation theologians, salvation consists in "God's gift of life and communion with God and neighbour." Sin is the absence of this fellowship, and so solidarity, which is the restoration of this fellowship, can be salvific. Ashworth also finds a correlate in those systematic theologians, such as Thomas Torrance and Kathryn Tanner, who propose a model of incarnation as atonement. God's solidarity with us in the incarnation,

enables us to become "at one" with God. See Ashworth, "How Divine Solidarity Liberates," *Scottish Journal of Theology* 72 (2019): 324–34, 326.

100. Dori Laub, "Bearing Witness or the Vicissitudes of Listening," in *Testimony: Crises of Witnessing in Literature, Psychoanalysis, and History*, by Shoshana Felman and Dori Laub (Abingdon: Taylor & Francis, 1992), 57–74, 57. Laub continues: the listener is "the blank screen on which the event comes to be inscribed for the first time."

101. Laub, "Bearing Witness or the Vicissitudes of Listening," 58. Jon Sobrino also describes witnessing as "a process of doing." See Sobrino, *Christ the Liberator*, 46.

102. Theologically, this resonates with the doctrine of *creatio ex nihilo*; the witness's testimony is created out of nothing. Just as creation comes to exist through Christ, so also it is Christ who is the creator of this witness to trauma.

103. This is another reason to prefer the terminology of witnessing over Jones's language of mirroring. Witnessing is active and creative, whereas mirroring is passive and static.

104. Richard Bauckham, "The Incarnation and the Cosmic Christ," in *Incarnation: On the Scope and Depth of Christology*, ed. Niels Henrik Gregersen (Minneapolis: Fortress Press, 2015), 25–58, 37–39; Niels Henrik Gregersen, "The Emotional Christ: Bonaventure and Deep Incarnation," *Dialog* 55 (2016): 247–61, 253–54.

105. As Gregory the Great states, "they [human beings] share existence with stones, like trees they are alive, like animals, they feel, and like the angels, they have understanding. If human beings, then, have something in common with every creature, in some sense human beings are every creature." See Gregory the Great, *Forty Gospel Homilies*, trans. Dom David Hurst, Monastic Studies Series (Piscataway, NJ: Gorgias Press, 2009), 227.

106. In Bonaventure's words, "since Christ, as a human being, has something from all of creation, and was transfigured, all is said to be transfigured in him." See Bonaventure, *The Sunday Sermons of St. Bonaventure*, ed. Robert F. Karris, trans. Timothy J. Johnson (St. Bonaventure, NY: Franciscan Institute, 2008), 217.

107. Gregersen, "The Cross of Christ in an Evolutionary World," 203. Moreover, "the microcosm of the sufferings of Jesus Christ—fully human and fully divine—may reveal God's presence in the sufferings and woes of creation at large." See Gregersen, "The Cross of Christ in an Evolutionary World," 195.

108. Bauckham, "The Incarnation and the Cosmic Christ," 43.

109. Gregersen, "Cur Deus Caro," 388.

110. This extension already occurs in the eucharist and in the church, but Graham Ward proposes that Christ's body "expands to embrace the whole of creation." See Ward, "The Displaced Body of Jesus Christ," 177.

111. Wallace, "The Wounded Spirit," 60. The same is true of Sallie McFague's work on the world as the body of God. See Sallie McFague, *Models of God: Theology for an Ecological, Nuclear Age* (London: SCM Press, 1987), 72.

5. The Rupture of Flesh: Deep Incarnation and Enfleshed Witnessing

1. Katie G. Cannon, "Womanist Perspectival Discourse and Cannon Formation," *Journal of Feminist Studies in Religion* 9 (1993): 29–37, 35.

2. Eva and Franco Mattes, *Fukushima Texture Pack*, Creative Capital (2016), accessed 11 November 2021, https://0100101110101101.org/fukushima-texture-pack/.

3. Eva and Franco Mattes, *Fukushima Texture Pack*, Creative Capital (2016), accessed 11 November 2021, https://0100101110101101.org/page/16/.

4. Serene Jones, *Trauma and Grace: Theology in a Ruptured World*, 2nd Edition (Louisville, KY: Westminster John Knox Press, 2019), 12.

5. Babette Rothschild, *The Body Remembers: The Psychophysiology of Trauma and Trauma Treatment* (New York: W. W. Norton, 2000), xiii; Bessel A. van der Kolk, *The Body Keeps the Score: Brain, Mind, and Body in the Healing of Trauma* (London: Penguin Books, 2015), 51–86.

6. Rothschild, *The Body Remembers*, 47; van der Kolk, *The Body Keeps the Score*, 2–3.

7. Rothschild, *The Body Remembers*, 7; van der Kolk, *The Body Keeps the Score*, 66.

8. Van der Kolk, *The Body Keeps the Score*, 86.

9. Cannon, "Womanist Perspectival Discourse and Cannon Formation," 35, quoted in van der Kolk, *The Body Keeps the Score*, 184.

10. Louis Heyse-Moore, "Does Gaia Experience Trauma?," *European Journal of Ecopsychology* 7 (2022): 75–99, 85. Note that Heyse-Moore discusses body memories in the context of a Gaian conception of ecological trauma. But the same still applies within my anthropomorphic conception of ecological trauma, as I discussed in Chapter 1.

11. Graham Ward, *The Politics of Discipleship: Becoming Postmaterial Citizens* (London: SCM Press, 2009), 222. For others, it is the body, and not the flesh, that is the primary "artifact produced for social control." See Mayra Rivera, *Poetics of the Flesh* (Durham, NC: Duke University Press, 2015), 7.

12. The body carries connotations of boundedness and integrity when in fact we do not really know where the body begins and ends. For example, work on biological mutualism reveals that the human body contains more bacterial cells than human cells. See Ron Sender et al., "Are We Really Vastly Outnumbered? Revisiting the Ratio of Bacterial to Host Cells in Humans," *Cell* 164 (2016): 337–40.

13. Paul J. Griffiths, *Christian Flesh* (Stanford, CA: Stanford University Press, 2018), 2.

14. Rivera, *Poetics of the Flesh*, 2.

15. Sharon V. Betcher, "Becoming Flesh of My Flesh: Feminist and Disability Theologies on the Edge of Posthumanist Discourse," *Journal of Feminist Studies in Religion*, 26 (2010): 107–18, 108.

16. Betcher, "Becoming Flesh of My Flesh," 107, 109.

17. Betcher, "Becoming Flesh of My Flesh," 108, 111.

18. Paul Voosen, "Ice Shelf Holding Back Keystone Antarctic Glacier Within Years of Failure," *Science* 374 (2021): 1420–21.

19. Mayra Rivera also argues, based on her reading of John's prologue, that it is not so easy to separate literal and figurative meanings of flesh; the two are intertwined. See Rivera, *Poetics of the Flesh*, 26.

20. Niels Henrik Gregersen, "The Idea of Deep Incarnation: Biblical and Patristic Resources," in *To Discern Creation in a Scattering World*, ed. Frederiek Depoortere and Jacques Haers (Leuven: Uitgeverij Peeters, 2013), 319–41, 322. Gregersen is guided in his interpretation of the Johannine prologue by the Copenhagen School of New Testament scholarship, which emphasizes the importance of Stoic philosophy (alongside Middle Platonic and Jewish influences) in shaping the writings of both John and Paul. See Gregersen, "God, Matter, and Information: Towards a Stoicizing Logos Christology," in *Information and the Nature of Reality: From Physics to Metaphysics*, ed. Paul Davies and Niels Henrik Gregersen (Cambridge: Cambridge University Press, 2010), 405–43. Stoic thought is germane to Gregersen's understanding of deep incarnation because it particularly emphasizes God's immanent material presence within creation. Hence, it is not hard for a Stoic to imagine *Logos* and *sarx* as coextensive. However, this reliance on Stoicism also has drawbacks for a theology of ecological trauma. Celia Deane-Drummond is critical of the Copenhagen School for overemphasizing the Hellenistic understanding of the cosmic elements of the *Logos* at the expense of the more historically focused Hebraic understanding. If deep incarnation aims to "ground Christology in earth processes," then a universal, Stoicizing Christology, says Deane-Drummond, does not help with this task. See Deane-Drummond, "Who on Earth Is Jesus Christ? Plumbing the Depths of Deep Incarnation," in *Christian Faith and the Earth: Current Paths and Emerging Horizons in Ecotheology*, ed. Ernst M. Conradie et al. (London: T & T Clark, 2014), 31–50, 33, 37; Deane-Drummond, "The Wisdom of Fools? A Theo-Dramatic Interpretation of Deep Incarnation," in *Incarnation: On the Scope and Depth of Christology*, ed. Niels Henrik Gregersen (Minneapolis: Fortress Press, 2015), 177–202, 179–80.

21. John 1:14. This verse is often translated as "the Word became flesh and tabernacled among us," referring to the portable dwelling place of God used by the Israelites. Yet the tent of the tabernacle was likely to have been made from animal skins. Therefore, to employ the metaphor that Christ "tabernacled among us" is to suggest that we should consider him entering animal flesh.

22. The closest it comes is in Philippians 2:7 where Christ is said to bear "human likeness." It is only at the councils of Nicaea and Chalcedon that "was incarnate" is supplemented by "was made man." See Niels Henrik Gregersen, "Deep Incarnation: Why Evolutionary Continuity Matters in Christology," *Toronto Journal of Theology* 26 (2010): 173–88, 174; Niels Henrik Gregersen, "Christology," in *Climate Change and Systematic Theology: Ecumenical Perspectives*, ed. Michael Northcott and Peter Scott (New York: Routledge, 2014), 33–50, 44.

23. Gregersen, "The Idea of Deep Incarnation," 329. Similarly, Denis Edwards

emphasizes that incarnation is far more than just the birth of Christ, it is the whole event of the Word becoming flesh. See Edwards, *Deep Incarnation: God's Redemptive Suffering with Creatures* (Maryknoll, NY: Orbis Books, 2019), xvii.

24. Niels Henrik Gregersen, "Cur Deus Caro: Jesus and the Cosmos Story," *Theology and Science* 11 (2013): 370–93, 370.

25. Gregersen, "Deep Incarnation: Why Evolutionary Continuity Matters in Christology," 181; Gregersen, "The Idea of Deep Incarnation," 328. However, Ronald Cole-Turner questions whether the Johannine concept of flesh is as inclusive as it first appears. For Cole-Turner, the semantic range of the Johannine *sarx* does not extend beyond physical lumps of tissue-filled flesh. See Cole-Turner, "Incarnation Deep and Wide: A Response to Niels Gregersen," *Theology and Science* 11 (2013): 424–35, 429. On the other hand, Joshua M. Moritz finds flesh to be "the most conceivably inclusive category of being and mode of existence known." See Moritz, "Deep Incarnation and the Imago Dei: The Cosmic Scope of the Incarnation in Light of the Messiah as the Renewed Adam," *Theology and Science* 11 (2013): 436–43, 440. Some authors have also proposed reading the Johannine flesh as referring specifically to warm-blooded animals. For example, Andrew Linzey writes that "the flesh assumed in the incarnation is not some hermetically sealed, tightly differentiated human flesh; it is the same organic flesh and blood which we share with other mammalian creatures," although this is not one of the ways that John uses *sarx* in his gospel. See Linzey, *Animal Rites: Liturgies of Animal Care* (London: SCM, 1999), 5. Note, too, how John's second definition of *sarx* as a reference to the sinful nature of humanity has often dominated Christian discussions of flesh. See Elizabeth A. Johnson, *Creation and the Cross: The Mercy of God for a Planet in Peril* (Maryknoll, NY: Orbis Books, 2018), 165–69. Mayra Rivera traces this link between sin and flesh back to Augustine and Paul. She also concurs with Gregersen's preference for the third interpretation of *sarx*, writing, "I assume flesh refers to all human flesh—and more." See Rivera, *Poetics of the Flesh*, 1, 19, 29–41.

26. Gregersen, "The Idea of Deep Incarnation," 321.

27. Elizabeth A. Johnson indicates that it is the decentering of humanity by contemporary science that prompts us to reexamine the multiple meanings of *sarx*. See Johnson, "Jesus and the Cosmos: Soundings in Deep Christology," in *Incarnation: On the Scope and Depth of Christology*, ed. Niels Henrik Gregersen (Minneapolis: Fortress Press, 2015), 133–56, 135–36.

28. Gregersen, "Deep Incarnation: Why Evolutionary Continuity Matters in Christology," 176.

29. Johnson, "Jesus and the Cosmos," 138.

30. Edwards, *Deep Incarnation*, 113.

31. See Gregersen, "Cur Deus Caro," 370; Niels Henrik Gregersen, "The Extended Body of Christ: Three Dimensions of Deep Incarnation," in *Incarnation: On the Scope and Depth of Christology*, ed. Niels Henrik Gregersen (Minneapolis: Fortress Press, 2015), 225–51, 225–26, quoted in Edwards, *Deep Incarnation*, 21.

32. Gregersen, "The Extended Body of Christ," 226. It is worth noting that the

relevance of deep incarnation to the seemingly hopeless corners of creation was part of Gregersen's motivation for formulating the concept in the first place. Gregersen was concerned with the problem of evolutionary theodicy, and so was trying to develop a Lutheran theology of the cross that would speak to the pain and suffering within the cosmos that goes beyond human sin. See Niels Henrik Gregersen, "The Cross of Christ in an Evolutionary World," *Dialog: A Journal of Theology* 40 (2001): 192–207, 192. In other words, deep incarnation is a direct "theological response to the pain, extinction, and death that are part of evolutionary emergence." See Edwards, *Deep Incarnation*, 1.

33. Gregersen, "The Cross of Christ in an Evolutionary World," 192, 197.

34. Elizabeth A. Johnson, *Ask the Beasts: Darwin and the God of Love* (London: Bloomsbury, 2014), 205; Johnson, *Creation and the Cross*, 188.

35. Johnson, *Creation and the Cross*, 178.

36. Johnson, *Ask the Beasts*, 192.

37. Gregersen, "Deep Incarnation: Why Evolutionary Continuity Matters in Christology," 176.

38. Gregersen, "Deep Incarnation: Why Evolutionary Continuity Matters in Christology," 182.

39. Johnson, *Ask the Beasts*, 203.

40. Niels Henrik Gregersen, "Introduction," in *Incarnation: On the Scope and Depth of Christology*, ed. Niels Henrik Gregersen (Minneapolis: Fortress Press, 2015), 1–21, 8.

41. See Chapter 4 and Shelly Rambo, *Spirit and Trauma: A Theology of Remaining* (Louisville, KY: Westminster John Knox Press, 2010), 37–44.

42. Jones, *Trauma and Grace*, 123.

43. Shelly Rambo, *Resurrecting Wounds: Living in the Afterlife of Trauma* (Waco, TX: Baylor University Press, 2017), 151.

44. Rambo, *Resurrecting Wounds*, 40.

45. Graham Ward, *Christ and Culture* (Malden, MA: Blackwell Publishers, 2005), 102.

46. Ward, *Christ and Culture*, 249.

47. Kelly Oliver, *Witnessing: Beyond Recognition* (Minneapolis: University of Minnesota Press, 2001), 86.

48. Oliver, *Witnessing: Beyond Recognition*, 88.

49. Oliver, *Witnessing: Beyond Recognition*, 90.

50. Van der Kolk, *The Body Keeps the Score*, 86. Erin Kidd offers another reformulation of the same phrase when she suggests that "the body of God keeps the score." See Kidd, "The Violation of God in the Body of the World: A Rahnerian Response to Trauma," *Modern Theology* 35 (2019): 663–82, 681.

51. Gregersen, "Introduction," 4.

52. Rosemary Radford Ruether, *Sexism and God-Talk* (London: SCM Press, 1983), 116.

53. Matthew Eaton, "Enfleshing Cosmos and Earth: An Ecological Theology

of Divine Incarnation," unpublished doctoral thesis (University of St. Michael's College, 2017), 87, accessed 6 November 2018, https://tspace.library.utoronto.ca/bitstream/1807/81412/6/Eaton_Matthew_201711_PhD_thesis.pdf.

54. Eaton, "Enfleshing Cosmos and Earth," 221n548.

55. As Laurel C. Schneider notes, God's incarnation as a human, and as a man, far from representing all flesh, often serves as a rationale for oppressing other forms of flesh. See Schneider, "Promiscuous Incarnation," in *The Embrace of Eros: Bodies, Desires, and Sexuality in Christianity*, ed. Margaret D. Kamitsuka (Minneapolis: Fortress Press, 2010), 231–46, 232. On the other hand, Duncan Reid offers one possible explanation when he writes, "the Word needed to become flesh, but then also specifically human flesh, not because humans are somehow better than all other flesh, but on the contrary, because humans are responsible for the corruption and suffering of the rest of creation." See Reid, "Enfleshing the Human: An Earth-Revealing, Earth-Healing Christology," in *Earth Revealing—Earth Healing: Ecology and Christian Theology*, ed. Denis Edwards (Collegeville, MN: The Liturgical Press, 2001), 69–83, 75.

56. For example, as Jürgen Moltmann says of humanity as microcosm, "this does not mean that all forms of life are present in the human being, but it does indicate the presence of all the elements of life." See Moltmann, "Is God Incarnate in All That Is?," in *Incarnation: On the Scope and Depth of Christology*, ed. Niels Henrik Gregersen (Minneapolis: Fortress Press, 2015), 119–32, 128.

57. As David S. Cunningham puts it, "God's incarnation is defined not so much by the accidental properties of this flesh (Jewish, male, human) as it is by its essential fleshly character." See Cunningham, "The Way of All Flesh: Rethinking the *Imago Dei*," in *Creaturely Theology: On God, Humans and Other Animals*, ed. Celia Deane-Drummond and David Clough (London: SCM Press, 2009), 100–117, 116.

58. Deane-Drummond, "The Wisdom of Fools?," 184. See also: Deane-Drummond, "Who on Earth Is Jesus Christ?," 40.

59. Norman Habel, "The Crucified Land: Towards Our Reconciliation With the Earth," *Colloquium* 28 (1996): 3–19, 14; my emphasis. This is a deliberate echo of Luther's statement about the eucharist, that "God is wholly in the grain and the grain is holy in God." Note too that for Habel flesh is synonymous with clay and earth.

60. James A. Nash, *Loving Nature: Ecological Integrity and Christian Responsibility* (Nashville, TN: Abingdon Press, 1991), 108. Nash's argument resonates here with the patristic notion of Christ as a microcosm of the macrocosm.

61. Nash, *Loving Nature*, 108–9.

62. Elizabeth Johnson expresses something akin to this idea when she says that we see a solidarity of Christ with humanity, a solidarity of God with Christ, and a solidarity of Christ with all of creation. See Johnson, *Creation and the Cross*, 159.

63. Elie Wiesel, *Night*, trans. Marion Wiesel (New York: Hill and Wang, 1958), 65. Wiesel, as a Jew, writes about God and not Christ. But Jürgen Moltmann quotes—and arguably ill-advisedly appropriates—Wiesel's line in a Christological

context in *The Crucified God*. See Moltmann, *The Crucified God: The Cross of Christ as the Foundation and Criticism of Christian Theology*, trans. R. A. Wilson and John Bowden (London: SCM Press, 1974), 274.

64. James H. Cone, *The Cross and the Lynching Tree* (Maryknoll, NY: Orbis Books, 2011), 158.

65. Cone, *The Cross and the Lynching Tree*, 165.

66. Cone, *The Cross and the Lynching Tree*, xv.

67. Habel, "The Crucified Land," 16.

68. Mark I. Wallace, *When God Was a Bird* (New York: Fordham University Press, 2019), 166. See Chapter 2 for further discussion.

69. Schneider, "Promiscuous Incarnation," 234.

70. Schneider, "Promiscuous Incarnation," 244.

71. Schneider, "Promiscuous Incarnation," 245.

72. Catherine Keller, *Intercarnations: Exercises in Theological Possibility* (New York: Fordham University Press, 2017), 1.

73. Keller, *Intercarnations*, 2.

74. Eaton, "Enfleshing Cosmos and Earth," 3.

75. Eaton, "Enfleshing Cosmos and Earth," 6.

76. Eaton, "Enfleshing Cosmos and Earth," 21.

77. Gregory Nazianzen, "The First Letter to Cledonius the Presbyter," in *On God and Christ: The Five Theological Orations and Two Letters to Cledonius*, trans. Lionel Wickham (Crestwood, NY: St. Vladimir's Seminary Press, 2002), 155–66, 158. A remarkably cognate sentiment can also be found in the book of Hebrews: "since, therefore, the children [of God] share flesh and blood, he himself likewise shared the same things. . . . Because he himself was tested by what he suffered, he is able to help those who are being tested." See Hebrews 2:14, 18, quoted in Gregersen, "The Extended Body of Christ," 240.

78. Gregersen, "Deep Incarnation: Why Evolutionary Continuity Matters in Christology," 184; Moritz, "Deep Incarnation and the Imago Dei," 441; Gregersen, "Christology," 45; Niels Henrik Gregersen, "Deep Incarnation: The Logos Became Flesh," in *Transformative Theological Perspectives*, ed. Karen L. Bloomquist (Minneapolis: Lutheran University Press, 2010), 167–81, 178; Edwards, *Deep Incarnation*, 84. Maurice F. Wiles points out that "healing" is a potentially unhelpful metaphor for salvation in this context. Is healing supposed to be physical, mental, spiritual, or something else altogether? See Wiles, "The Unassumed Is the Unhealed," *Religious Studies* 4 (1968): 47–56, 51.

79. Gregersen, "The Idea of Deep Incarnation," 335.

80. Wiles, "The Unassumed Is the Unhealed," 48.

81. Gregory is unlikely to have had the rest of creation in mind when he wrote about the unassumed and the unhealed. These subsequent interpretations are therefore a development of the tradition.

82. Karl Rahner, *Mission and Grace: Essays in Pastoral Theology II*, trans. Cecily Hastings and Richard Strachan (London: Sheed & Ward, 1963–1966), 39–42, quoted

in Edwards, *Deep Incarnation*, 85. Gregersen refers to this as a "twofold assumption": God assumes flesh in the incarnation so that all flesh is assumed into the divine life at the eschaton. In contrast to the proponents of pancarnation, Gregersen only imagines that Christ is incarnate in everything eschatologically. See Niels Henrik Gregersen, "The Twofold Assumption: A Response to Cole-Turner, Moritz, Peters and Peterson," *Theology and Science* 11 (2013): 455–68, 455–56.

83. Gregersen, "Introduction," 2.

84. These modes of presence include theophanies, visions, encounters, addresses, conversations, inspirations, empowerments, providential care, and sacraments. See Richard Bauckham, "The Incarnation and the Cosmic Christ," in *Incarnation: On the Scope and Depth of Christology*, ed. Niels Henrik Gregersen (Minneapolis: Fortress Press, 2015), 25–58, 27.

85. Christopher Southgate, "Depth, Sign and Destiny: Thoughts on Incarnation," in *Incarnation: On the Scope and Depth of Christology*, ed. Niels Henrik Gregersen (Minneapolis: Fortress Press, 2015), 203–24, 207.

86. For example, "the identity of God as Love can't be revealed in a tomato or in a mussel, nor in the birth and decay of stars and galaxies." See Gregersen, "The Twofold Assumption," 458. Unlike Bauckham, Gregersen does still insist on God being internally, rather than merely externally, present. As he writes, "the embodied Word of God shares *from within* the sufferings of all who suffer from the power of tsunamis, earthquakes, and hunger, and *takes the side* of the victims of the horrors that human beings inflict upon one [an]other." See Gregersen, "The Extended Body of Christ," 235.

87. Celia Deane-Drummond, "Deep Incarnation Between Balthasar and Bulgakov: The Form of Beauty and the Wisdom of God," in *Envisioning the Cosmic Body of Christ*, ed. Aurica Jax and Saskia Wendel (Abingdon: Routledge, 2019), 101–13, 110. Similarly, Jürgen Moltmann affirms that God is only incarnate in all that is at the eschaton. See Moltmann, "Is God Incarnate in All That Is?," 119, 131.

88. Deane-Drummond, "Deep Incarnation Between Balthasar and Bulgakov," 111.

89. Gregersen, "Cur Deus Caro," 388. Gregersen is also keen to stress that there should not be a strong divide between Christ's person and his works, that is, between his being and his acts, or his ontology and his agency. Such a distinction, he says, is a hangover from an outdated Aristotelian substance metaphysics. See Niels Henrik Gregersen, "Deep Incarnation: Opportunities and Challenges," in *Incarnation: On the Scope and Depth of Christology*, ed. Niels Henrik Gregersen (Minneapolis: Fortress Press, 2015), 361–79, 367–68.

90. See Gregersen, "Deep Incarnation: Why Evolutionary Continuity Matters in Christology," 187n26.

91. Ernst M. Conradie, "Resurrection, Finitude, and Ecology," in *Resurrection: Theological and Scientific Assessments*, ed. Ted Peters et al. (Grand Rapids, MI: Eerdmans, 2002), 277–96, 292.

92. Conradie, "Resurrection, Finitude, and Ecology," 292. Stefan Skrimshire

glosses Conradie's account of material inscription as a way of describing "God's memory of the earth." See Skrimshire, "Anthropocene Fever: Memory and the Planetary Archive," in *Religion and the Anthropocene*, ed. Celia Deane-Drummond et al. (Eugene, OR: Cascade Books, 2017), 138–54, 152.

93. Conradie, "Resurrection, Finitude, and Ecology," 292.

94. Conradie, "Resurrection, Finitude, and Ecology," 293. See also: Denis Edwards, "Every Sparrow That Falls to the Ground: The Cost of Evolution and the Christ-Event," *Ecotheology* 11 (2006): 103–23, 117.

95. Gregersen, "The Cross of Christ in an Evolutionary World," 203.

96. Conradie, "Resurrection, Finitude, and Ecology," 293.

97. Gregersen writes, "deep incarnation . . . suggests that God not only tolerates material existence, but also accepts it and incorporates it in a divine embrace . . . the incarnate God incorporates the physical realm of creation—not only its lawlike or 'logical' aspects (as highlighted already in Patristic writers), but also the chaotic aspects of creation." See Gregersen, "Cur Deus Caro," 375.

98. Cannon, "Womanist Perspectival Discourse and Cannon Formation," 35.

99. It is important to note that Christ's flesh is marked, and therefore serves as a witness, during his life as well as his death. Christ's circumcision is a particularly clear example of this fleshly marking. See Luke 1:59; Ward, *Christ and Culture*, 256.

6. The Rupture of Time: Witnessing Anthropocene Scars

1. James Hutton, *Theory of the Earth; or an Investigation of the Laws Observable in the Composition, Dissolution, and Restoration of Land Upon the Globe* (*Transactions of the Royal Society of Edinburgh*) (London: Forgotten Books, [1788] 2007), 96.

2. Anil Narine, *Eco-Trauma Cinema* (New York: Routledge, 2015), 13.

3. Hutton, *Theory of the Earth*, 96.

4. The specific phrase "deep time" was coined more recently by John McPhee. McPhee writes, "numbers do not seem to work well with regard to deep time. Any number above a couple of thousand years—fifty thousand, fifty million—will with nearly equal effect awe the imagination to the point of paralysis." See McPhee, *Basin and Range* (New York: Farrar, Straus and Giroux, 1981), 20.

5. Cymene Howe, "'Timely' Theorizing the Contemporary," *Fieldsights* (2016), accessed 5 March 2021, https://culanth.org/fieldsights/timely; Dipesh Chakrabarty, "Humanities in the Anthropocene: The Crisis of an Enduring Kantian Fable," *New Literary History* 47 (2016): 377–97; Franklin Ginn et al., "Introduction: Unexpected Encounters with Deep Time," *Environmental Humanities* 10 (2018): 213–25.

6. Paul J. Crutzen and Eugene F. Stoermer, "The 'Anthropocene,'" *International Geosphere-Biosphere Programme Global Change Newsletter* (2000), 17–18; Paul J. Crutzen, "Geology of Mankind," *Nature* 415 (2002), 23. In 2024, scientists from the International Union of Geological Sciences officially rejected the Anthropocene as a new epoch on the geological timescale, although they did recognize the term as

an "invaluable descriptor" that will continue to be widely used. See International Union of Geological Sciences, "The Anthropocene" (2024), accessed 6 August 2024, https://www.iugs.org/_files/ugd/f1fc07_40d1a7ed58de458c9f8f24de5e739663 .pdf?index=true. Much of the debate concerns the precise timing of the beginning of the Anthropocene. See Simon L. Lewis and Mark A. Maslin, "Defining the Anthropocene," *Nature* 519 (2015): 171–80. There is also concern that the concept implies that all of humanity is equally responsible for ecological devastation when that is manifestly not the case. See Andreas Malm and Alf Hornborg, "The Geology of Mankind? A Critique of the Anthropocene Narrative," *Anthropocene Review* 1 (2014): 62–69. Nevertheless, usage of the word is so widespread that the Anthropocene remains a helpful term for describing the extent of human impacts on the planet, and I continue to use it here.

7. Karen O'Donnell, *Broken Bodies: The Eucharist, Mary and the Body in Trauma Theology* (London: SCM Press, 2018), 7; Shelly Rambo, *Spirit and Trauma: A Theology of Remaining* (Louisville, KY: Westminster John Knox Press, 2010), 19.

8. Cathy Caruth, "Trauma and Experience: Introduction," in *Trauma: Explorations in Memory*, ed. Cathy Caruth (Baltimore, MD: Johns Hopkins University Press, 1995), 3–12, 4.

9. Caruth, "Trauma and Experience: Introduction," 7.

10. Given these recurrences, another common symptom of trauma is a continuing state of hyperarousal, or a "persistent expectation of danger." Trauma survivors are on edge about the possibility of future violence and a variety of stimuli can trigger further flashbacks. See Judith Lewis Herman, *Trauma and Recovery* (New York: Basic Books, 1992), 35.

11. Herman, *Trauma and Recovery*, 35.

12. Rambo, *Spirit and Trauma*, 19. As Flora A. Keshgegian writes, "if we can try not to solve or 'heal' time and its wounds and if we do not cover them over with worn narratives as bandages, then the fractured time sense of the traumatized may serve to correct traditional, linear views." See Keshgegian, *Time for Hope: Practices for Living in Today's World* (New York: Continuum, 2006), 121.

13. See Chapter 3 for further discussion of Hurricane Katrina.

14. The IPCC states that it is "likely" that the proportion of tropical cyclones (hurricanes) reaching Category 3–5 has increased over the last four decades, and they have "high confidence" that climate change has increased the intensity of precipitation associated with these events. It is also "very likely" that the frequency and intensity of these events will increase with ongoing global warming. See IPCC, "Summary for Policymakers," in *Climate Change 2021: The Physical Science Basis. Contribution of Working Group I to the Sixth Assessment Report of the Intergovernmental Panel on Climate Change*, ed. V. Masson-Delmotte et al. (Cambridge: Cambridge University Press, 2021), 9, 16.

15. Stefan Skrimshire, "Deep Time and Secular Time: A Critique of the Environmental 'Long View,'" *Theory, Culture and Society* 36 (2019): 63–81, 66.

16. Will Steffen et al., "Trajectories of the Earth System in the Anthropocene,"

PNAS 115 (2018): 8252–59; Timothy Lenton et al., "Climate Tipping Points—Too Risky to Bet Against," *Nature* 575 (2019): 592–95.

17. Stefan Skrimshire, "Eternal Return of the Apocalypse," in *Future Ethics: Climate Change and Apocalyptic Imagination*, ed. Stefan Skrimshire (London: Continuum, 2010), 228.

18. Rob Nixon, *Slow Violence and the Environmentalism of the Poor* (Cambridge, MA: Harvard University Press, 2011), 2.

19. Nixon, *Slow Violence and the Environmentalism of the Poor*, 2.

20. Nixon, *Slow Violence and the Environmentalism of the Poor*, 6.

21. The Anthropocene Working Group, the international body of scientists who were charged with determining the geological existence of the Anthropocene, write that many of the changes in the Anthropocene "will persist for millennia or longer, and are altering the trajectory of the Earth System, some with permanent effect." See Anthropocene Working Group, "What Is the Anthropocene? – Current Definition and Status," *Subcommission on Quaternary Stratigraphy*, accessed 12 January 2022, https://quaternary.stratigraphy.org/working-groups/anthropocene.

22. Clive Hamilton, *Defiant Earth: The Fate of Humans in the Anthropocene* (Cambridge: Polity Press, 2017), 7–8.

23. David Archer, "Fate of Fossil Fuel CO_2 in Geologic Time," *Journal of Geophysical Research C: Oceans* 110 (2005): 1–6; J. E. N. Veron, "Mass Extinctions and Ocean Acidification: Biological Constraints on Geological Dilemmas," *Coral Reefs* 27 (2008): 459–72, quoted in Richard W. Miller, "Deep Responsibility for the Deep Future," *Theological Studies* 77 (2016): 436–65, 441–42.

24. Skrimshire, "Deep Time and Secular Time," 66.

25. As Bronislaw Szerszynski writes, "memories laid down in rock are like a great archive." See Szerszynski, "How the Earth Remembers and Forgets," in *Political Geology: Active Stratigraphies and the Making of Life*, ed. Adam Bobbette and Amy Donovan (Cham, Switzerland: Palgrave Macmillan, 2018), 219–36, 232. The Earth's "memories" are also themselves a form of witnessing. According to Michael Richardson, "geological formations witness the passage of deep time, the arrival and departure of ice ages, the life and death of forests, and the passage of animal life." For Richardson, this nonhuman witnessing is especially important because it not only communicates forms of violence that are otherwise occluded, but also reckons with the scale and complexity of ecological catastrophe. See Richardson, *Nonhuman Witnessing: War, Data, and Ecology After the End of the World* (Durham, NC: Duke University Press, 2024), 8, 14, 117.

26. Herman, *Trauma and Recovery*, 35.

27. Hamilton, *Defiant Earth*, 3.

28. Clive Hamilton and Jacques Grinevald, "Was the Anthropocene Anticipated?," *Anthropocene Review* 2 (2015): 59–72, 66–67.

29. Clive Hamilton, "The Anthropocene as Rupture," *Anthropocene Review* 3 (2016): 93–106, 100.

30. Hamilton, *Defiant Earth*, 3.

31. Marcia Bjornerud, *Timefulness: How Thinking Like a Geologist Can Help Save the World* (Princeton, NJ: Princeton University Press, 2018), 130.

32. Christina Fredengren, "Re-Wilding the Environmental Humanities: A Deep Time Comment," *Current Swedish Archaeology* 26 (2018): 50–60, 53.

33. See Szerszynski, "How the Earth Remembers and Forgets," 233.

34. Richard Irvine, *An Anthropology of Deep Time: Geological Temporality and Social Life*, New Departures in Anthropology (Cambridge: Cambridge University Press, 2020), 10.

35. Irvine, *An Anthropology of Deep Time*, 99.

36. Irvine, *An Anthropology of Deep Time*, 106.

37. Rambo, *Spirit and Trauma*, 20. Furthermore, Augustine's well-known claim that the past does not exist, except in the form of memories of the past in the present, does not do justice to the temporal rupture of trauma. When a survivor experiences a flashback, the past is relived as present; the past exists as an ongoing wounding that is stored in the flesh. As Keshgegian writes, "when someone is in this trauma time, the traumatic past is present not only as memory but as event, as 'reality.'" See Augustine, *Confessions*, trans. Henry Chadwick (Oxford: Oxford University Press, 1998), 233–35; Keshgegian, *Time for Hope*, 100.

38. The same might also be said of the wider Christian metanarrative of fall, cross, and redemption. A good example of this purely linear understanding was proposed by Oscar Cullmann in the 1940s. He suggests that "revelation and salvation take place along the course of an ascending time line," and claims that there is a "strictly straight-line conception of time in the New Testament." See Cullmann, *Christ and Time: The Primitive Christian Conception of Time and History*, trans. Floyd V. Filson (London: SCM Press, 1951), 32.

39. Rambo, *Spirit and Trauma*, 143. Sharon V. Betcher refers to this as the "ontotheological history of creation-fall-redemption." See Betcher, *Spirit and the Politics of Disablement* (Minneapolis: Fortress Press, 2007), 43.

40. Rambo, *Spirit and Trauma*, 128. In fact, the allure of redemptive motifs is so powerful that they frequently recur in secular culture. Phrases such as "no pain, no gain" or "every cloud has a silver lining" are predicated on a narrative of growth that elides the reality of suffering. See Rambo, *Spirit and Trauma*, 146. In these secularized versions of linear salvation history paradise becomes possible on Earth through the ingenuity of humanity alone. See Keshgegian, *Time for Hope*, 35.

41. Rambo, *Spirit and Trauma*, 143.

42. Rambo, *Spirit and Trauma*, 127.

43. Rambo, *Spirit and Trauma*, 50n15.

44. Rambo, *Spirit and Trauma*, 127.

45. Hans Urs von Balthasar, *Heart of the World*, trans. Erasmo S. Leiva (San Francisco: Ignatius Press, 1979), 157.

46. Rambo says comparatively little about eschatology, but she does provide hints that it is another area of doctrinal reflection, alongside Christology, that needs to be reconceived in light of the traumatic rupture of time. In the context

of human warfare, she writes the following: "Within the framework of Christian theology, concepts of eschatology could be re-approached in terms of the dynamics of 'ongoingness' and disrupted temporality that are represented in war trauma. . . . It is equally important to note that eschatological visions might also be operative in fueling war. For example, religious vocabulary of the afterlife may provide visions of eternal reign, the Kingdom of God, and final judgment that can serve to sanction and justify violence in the present in the name of a more ultimate religious reality." See Shelly Rambo, "Changing the Conversation: Theologizing War in the Twenty-First Century," *Theology Today*, 69 (2013): 441–62, 450.

47. Rambo, *Spirit and Trauma*, 50n15.

48. Hans Urs von Balthasar, *First Glance at Adrienne von Speyr*, trans. Antje Lawry and Sergia Englund (San Francisco: Ignatius Press, 1981), 35, 64. The possibilities of a spatial theology are intriguing and have been taken up more fully by Vítor Westhelle. See Westhelle, *Eschatology and Space: The Lost Dimension in Theology Past and Present* (New York: Palgrave Macmillan, 2012).

49. Shelly Rambo, *Resurrecting Wounds: Living in the Afterlife of Trauma* (Waco, TX: Baylor University Press, 2017), 7.

50. Rambo, *Resurrecting Wounds*, 149.

51. Rambo, *Resurrecting Wounds*, 5.

52. Rambo, *Resurrecting Wounds*, 6.

53. As Rob Nixon writes, "the deep-time thinking that celebrates natural healing is strategically disastrous if it provides political cover for reckless corporate short-termism." See Nixon, *Slow Violence and the Environmentalism of the Poor*, 21–22.

54. Niels Henrik Gregersen, "The Extended Body of Christ: Three Dimensions of Deep Incarnation," in *Incarnation: On the Scope and Depth of Christology*, ed. Niels Henrik Gregersen (Minneapolis: Fortress Press, 2015), 225–51, 243; my emphasis.

55. As Whitney A. Bauman notes, the imposition of chronological time on the planetary community is an illusion—one that is only strengthened by the fossil-fueled acceleration of time. See Bauman, "Sourdough Time and the Time of Protest: Reflections on the Pace of Planetary Becoming," *Minding Nature* 13 (2020): 94–99, 95.

56. By contrast, the later Paul—or an early follower of Paul—writes in the letter to the Colossians that "you have been raised with Christ," upending the progression just asserted. The present perfect verb form allows for the resurrection of the followers of Christ as an ongoing phenomenon—past, present, and future—rather than merely as a deferred reality. See Gregersen, "The Extended Body of Christ," 243–44; Colossians 3:1. This is echoed by Chris Tilling, who writes that "Paul's Christ seems to disrupt those sequential notions of narrative time that undergird . . . certain telic narrative approaches." He continues, "for Paul time bends around Christ such that the past story of Christ is the present." See Tilling, "Paul, Christ, and Narrative Time," in *Christ and the Created Order*, ed. Andrew B. Torrance and Thomas H. McCall (Grand Rapids, MI: Zondervan, 2018), 151–66, 157, 161.

57. Gregersen, "The Extended Body of Christ," 243.
58. Gregersen, "The Extended Body of Christ," 250.
59. Gregersen, "The Extended Body of Christ," 227–28.
60. Gregersen, "The Extended Body of Christ," 251.
61. Niels Henrik Gregersen, "Deep Incarnation: Why Evolutionary Continuity Matters in Christology," *Toronto Journal of Theology* 26 (2010): 173–88, 173.
62. Keshgegian, *Time for Hope*, 22–23.
63. Keshgegian, *Time for Hope*, 30. In addition, Keshgegian outlines at least two further conceptions of time: apocalyptic time, where a linear narrative is punctuated by dramatic divine interventions; and eternal time, as seen for example in the wisdom literature, where changelessness is recognized as a norm. In the book of Ecclesiastes, it is declared that "what has been is what will be, and what has been done is what will be done; there is nothing new under the sun." See Ecclesiastes 1:9. Christian theologians have often employed the Greek distinction between *chronos* and *kairos*. The former refers to the chronological clock time that Rambo and Gregersen find so problematic, while the latter signifies something more like an opportune moment for action: the proper setting for a particular phenomenon, whether that be birth, death, or even the incarnation. Again, Ecclesiastes provides a good example of kairological time: "For everything there is a season, and a time for every matter under heaven: a time to be born, and a time to die; a time to plant, and a time to pluck up what is planted; a time to kill, and a time to heal; a time to break down, and a time to build up [etc.]." See Ecclesiastes 3:1–8. Time here is translated as *kairos* in the Septuagint. *Kairos* events would not have been possible under different conditions or in different contexts. See John E. Smith, "Time, Times, and the 'Right Time'; Chronos and Kairos," *The Monist* 53 (1969): 1–13; Jürgen Moltmann, *God in Creation: An Ecological Doctrine of Creation*, trans. Margaret Kohl (London: SCM Press, 1985), 118; Ernst M. Conradie, "Resurrection, Finitude, and Ecology," in *Resurrection: Theological and Scientific Assessments*, ed. Ted Peters et al. (Grand Rapids, MI: Eerdmans, 2002), 277–96, 282; Michael S. Northcott, "Eschatology in the Anthropocene: From the Chronos of Deep Time to the Kairos of the Age of Humans," in *The Anthropocene and the Global Environment Crisis*, ed. Clive Hamilton et al. (Abingdon: Routledge, 2015), 100–111, 107–9.
64. Catherine Keller, *On the Mystery: Discerning Divinity in Process* (Minneapolis: Fortress Press, 2008), 174, quoted in Rambo, *Spirit and Trauma*, 126.
65. Rambo, *Spirit and Trauma*, 127.
66. Note how revisiting, repeating, recombining, and reinterpreting have a different quality to rebuilding, restoring, recovering, and resurrecting.
67. Rambo, *Spirit and Trauma*, 127.
68. Keller, *On the Mystery*, 161. Keller is consequently highly critical of creation *ex nihilo*, writing that "Beginning is going on. Everywhere. Amidst all the endings, so rarely ripe or ready." See Catherine Keller, *Face of the Deep: A Theology of Becoming* (Abingdon: Routledge, 2003), 3.
69. Keller, *On the Mystery*, 170.

70. For example, Robert W. Jenson writes, "time, in any construal adequate to the gospel, does not in fact march in this wooden fashion. Time, as we see it framing biblical narrative, is neither linear nor cyclical but perhaps more like a helix, and what it spirals around is the risen Christ." See Jenson, "Scripture's Authority in the Church," in *The Art of Reading Scripture*, ed. Ellen F. Davis and Richard B. Hays (Grand Rapids, MI: Eerdmans, 2003), 27–37, 35, quoted in Tilling, "Paul, Christ, and Narrative Time," 161.

71. Neither Keller nor Rambo pursue helical time in a Christological key because they remain nervous about heavy-handed Christologies that appear to promote self-sacrifice or impose redemption without remaining open to the realities of ongoing suffering and change. But Serene Jones references recapitulation as an important component of her idea of the "mirrored cross." See Chapter 4 and Jones, *Trauma and Grace*, 77.

72. Ephesians 1:10 also says that all things will be recapitulated in Christ.

73. Irenaeus writes, "and just as through a disobedient virgin man was struck and, falling, died, so also by means of a virgin, who obeyed the word of God, man, being revivified, received life. . . . For it was necessary for Adam to be recapitulated in Christ, that 'mortality might be swallowed up in immortality'; and Eve in Mary, that a virgin, become an advocate for a virgin, might undo and destroy the virginal disobedience by virginal obedience." See Irenaeus, *On the Apostolic Preaching*, trans. John Behr (Crestwood, NY: St. Vladimir's Seminary Press, 1997), §33.

74. Again, Irenaeus writes, "and the transgression which occurred through the tree was undone by the obedience of the tree—which [was shown when] the Son of Man, obeying God, was nailed to the tree." See Irenaeus, *On the Apostolic Preaching*, §33.

75. Irenaeus writes, "and since He is the Word of God Almighty, who invisibly pervades the whole creation, and encompasses its length, breadth, height and depth—for by the Word of God everything is administered—so too was the Son of God crucified in these [fourfold dimensions], having been imprinted in the form of the cross in everything." Irenaeus, *On the Apostolic Preaching*, §34. According to Denis Minns, Irenaeus derives this idea from Plato's *Timaeus*, via Justin Martyr's *First Apology*. Specifically, Plato references the world soul being arranged in the form of the Greek letter X, which Justin takes as a reference to the cross of Christ imprinted on the universe. See Minns, *Irenaeus: An Introduction* (London: T & T Clark, 2010), 109. On the cruciform creation, see Holmes Rolston III, *Science & Religion: A Critical Survey* (Philadelphia: Templeton Foundation Press, 2006), 133–46, 286–92; Holmes Rolston III, *Genes, Genesis and God: Values and Their Origins in Natural and Human History* (Cambridge: Cambridge University Press, 1999), 303–6; Holmes Rolston III, "Redeeming a Cruciform Nature," *Zygon* 53 (2018): 739–51. Minns hints at the same parallel between Irenaeus and Rolston when he writes that the cross "mirrors the cruciform stamp of his [Christ's] presence in the universe." See Minns, *Irenaeus*, 109. Similarly, John Behr highlights how Christ,

the Word of God, creates in a "cruciform manner." See Behr, *Irenaeus of Lyons: Identifying Christianity* (Oxford: Oxford University Press, 2013), 135.

76. Denis Edwards, *Deep Incarnation: God's Redemptive Suffering with Creatures* (Maryknoll, NY: Orbis Books, 2019), 122.

77. Keller, *Face of the Deep*, 221.

78. As Edwards puts it, "Adam's bodily humanity is shaped according to Christ's bodily humanity." See Edwards, *Deep Incarnation*, 36. And Denis Minns writes, "Adam was consequent on Christ, and not the other way around." See Minns, *Irenaeus*, 100.

79. Matthew C. Steenberg writes, for example, that "there is a definable, discernible *telos* toward which the creation is moving" and that "it is an entity in motion, in advancement." See Steenberg, *Irenaeus on Creation: The Cosmic Christ and the Saga of Redemption* (Leiden: Brill, 2008), 52.

80. Irenaeus, *On the Apostolic Preaching*, §31.

81. Edwards, *Deep Incarnation*, 53. Elsewhere, Irenaeus claims that "all things have been created for the service of man," and that "when the creation is restored, all the animals should obey and be in subjection to man." See Irenaeus, "Against Heresies," in *Ante-Nicene Fathers, Vol. 1*, ed. Alexander Roberts et al., trans. Alexander Roberts and William Rambaut (Buffalo, NY: Christian Literature Publishing Co., 1885), V.14.1 & V.33.4.

82. Niels Henrik Gregersen, "Deep Incarnation and Kenosis: In, With, Under, and As: A Response to Ted Peters," *Dialog* 52 (2013): 251–62, 256.

83. Niels Henrik Gregersen, "The Extended Body: The Social Body of Jesus According to Luke," *Dialog* 51 (2012): 234–44, 236; Niels Henrik Gregersen, "Cur Deus Caro: Jesus and the Cosmos Story," *Theology and Science* 11 (2013): 370–93, 380; Matthew 1:1–16; Luke 3:23–38.

84. Richard Irvine, "Seeing Environmental Violence in Deep Time," *Environmental Humanities* 10 (2018): 257–72, 259.

85. The human activities of the last two to three millennia, including the incarnation of Christ, can and should be embedded within a longer narrative about human origins. But deep time goes a step further than deep history, setting this story of human origins within the context of planetary time. See Niels Henrik Gregersen, "Deep Incarnation: From Deep History to Post-Axial Religion," *HTS Teologiese Studies/Theological Studies* 72 (2016): 1–12, 7; Gregersen, "Cur Deus Caro," 376; Celia Deane-Drummond, "Performing the Beginning in the End: A Theological Anthropology for the Anthropocene," in *Religion and the Anthropocene*, ed. Celia Deane-Drummond et al. (Eugene, OR: Cascade Books, 2017), 173–87, 175. However, if one understands Adam to mean "from the Earth" (*adamah*), then the genealogy does suggest that we think of Christ as "the son of the son of the Earth," establishing a possible link between Christ and a deep earthly time. See Niels Henrik Gregersen, "The Idea of Deep Incarnation: Biblical and Patristic Resources," in *To Discern Creation in a Scattering World*, ed. Frederiek Depoortere and Jacques Haers (Leuven: Uitgeverij Peeters, 2013), 319–41, 329.

86. John 1:3.
87. Rambo, *Resurrecting Wounds*, 5.
88. Rambo, *Spirit and Trauma*, 104.
89. Rambo, *Spirit and Trauma*, 104.
90. Luke 24:39.
91. John 20:20.
92. John 20:25.
93. Peter Widdicombe, "The Wounds and the Ascended Body: The Marks of Crucifixion in the Glorified Christ from Justin Martyr to John Calvin," *Laval Théologique et Philosophique* 59 (2003): 137–54, 137–38.
94. John M. Hull, "The Broken Body in a Broken World: A Contribution to a Christian Doctrine of the Person from a Disabled Point of View," *Journal of Religion, Disability & Health* 7 (2004): 5–23, 20; Widdicombe, "The Wounds and the Ascended Body," 148.
95. Quoted in Hull, "The Broken Body in a Broken World," 16–17.
96. John Hull is keen to stress that we are not merely talking about surface scars but wounds that are still open and deep enough for Thomas to insert his hand. Caravaggio's famous depiction of *The Incredulity of St. Thomas* emphasizes the depth to which Thomas can insert his finger into Christ's flesh. Candida Moss disagrees. She prefers to interpret the marks on Christ's body as scars rather than wounds, partly because "scars were the ultimate form of identification in the ancient world," and partly because scars would indicate that Christ's body had undergone a natural process of healing, further reinforcing its physicality. Meanwhile, Maja Whitaker argues against Moss, backing up Hull's initial view. For Whitaker, "wound" rather than "scar" is the better translation of John's gospel, and contemporary medicine suggests that Christ's wounds would not have healed over in the time between his crucifixion and his encounter with Thomas. See Hull, "The Broken Body in a Broken World," 18; Candida R. Moss, "The Marks of the Nails: Scars, Wounds and the Resurrection of Jesus in John," *Early Christianity* 8 (2017): 48–68, 58–65; Maja Whitaker, "The Wounds of the Risen Christ: Evidence for the Retention of Disabling Conditions in the Resurrection Body," *Journal of Disability and Religion* (2021): 1–14, 2–6.
97. Martin Luther, "A Meditation on Christ's Passion," in *The Works of Martin Luther, Volume 42: Devotional Writings, Volume One*, ed. Martin O. Dietrich, trans. Martin H. Bertram (Charlottesville, VA: Fortress Press, 1969), 3–14, 12; John Calvin, *Commentary on the Gospel According to John, Volume 2*, trans. William Pringle (Grand Rapids, MI: Eerdmans, 1956), 265.
98. Rambo, *Resurrecting Wounds*, 20–29.
99. Rambo, *Resurrecting Wounds*, 32.
100. Widdicombe, "The Wounds and the Ascended Body," 153.
101. Conradie, "Resurrection, Finitude, and Ecology," 293. Rambo voices the same concern when she writes that "the realities of the ongoing crosses of history, where wounds return and surface, could lead to a belief that ongoing violence

is inevitable." See Rambo, *Resurrecting Wounds*, 94. But Rambo stresses that we should not be allowing ourselves to permit violence as the status quo. Rather, the persistence of past wounds should serve as a permanent stimulus to try to prevent future violence.

102. Matthew Eaton, "Conclusion: Ecocide as Deicide: Eschatological Lamentation and the Possibility of Hope," in *Integral Ecology for a More Sustainable World: Dialogues with Laudato Si'*, ed. Dennis O'Hara et al. (Lanham, MD: Lexington Books, 2019), 359.

103. Zechariah 12:10.

104. Revelation 13:8.

105. Hull, "The Broken Body in a Broken World," 21. See also: Susanne Rappmann, "The Disabled Body of Christ as a Critical Metaphor—Towards a Theory," *Journal of Religion, Disability & Health* 7 (2004): 25–40; Deborah Beth Creamer, "Including All Bodies in the Body of God: Disability and the Theology of Sallie McFague," *Journal of Religion, Disability & Health* 9 (2006): 55–69.

106. Danielle Tumminio Hansen, "Remembering Rape in Heaven: A Constructive Proposal for Memory and the Eschatological Self," *Modern Theology* 37 (2021): 662–78.

107. There is also a further trinitarian question to be addressed here, namely, how Christ's wounded body affects a doctrine of God. As Peter Widdicombe expresses it, "the enduring reality of the wounds testify to the intimate connection between the economy of God's salvific work within the created order and the eternal economy." See Widdicombe, "The Wounds and the Ascended Body," 137. Theologians such as Jürgen Moltmann and Paul Fiddes already understand the crucifixion as a trinitarian event. As Moltmann says, "the theology of the cross must be the doctrine of the Trinity and the doctrine of the Trinity must be the theology of the cross." His point is that Christ's suffering, wounding, and death is not to be addressed via a Christological doctrine of Christ's two natures (where the human nature is passible, and the divine nature is impassible). Instead, the whole Trinity is already involved: "the Son suffers dying, the Father suffers the death of the Son." If suffering is not alien to the internal dynamics of the Trinity (Moltmann follows Rahner in refusing a distinction between the immanent and the economic Trinity), then Christ's persistent wounds do not pose any great difficulty for this doctrine of God. See Jürgen Moltmann, *The Crucified God: The Cross of Christ as the Foundation and Criticism of Christian Theology*, trans. R. A. Wilson and John Bowden (London: SCM Press, 1974), 241, 243. Similarly, Fiddes makes clear that "God suffers as Trinity." He continues, "within the divine perichoresis, all three persons suffer, but in different ways according to the distinction of relations." See Paul S. Fiddes, *Participating in God: A Pastoral Doctrine of the Trinity* (London: Darton, Longman & Todd, 2000), 179, 184.

108. Bruno Latour, *Facing Gaia: Eight Lectures on the New Climatic Regime*, trans. Catherine Porter (Cambridge: Polity Press, 2017), 118; Stefan Skrimshire, "Anthropocene Fever: Memory and the Planetary Archive," in *Religion and the*

Anthropocene, ed. Celia Deane-Drummond et al. (Eugene, OR: Cascade Books, 2017), 138–54, 139. Keller dubs this second possibility the "crash scene of the Anthropocene." See Catherine Keller, *Political Theology of the Earth: Our Planetary Emergency and the Struggle for a New Public* (New York: Columbia University Press, 2018), 71.

109. The idea that the rock record speaks to future inhabitants of the Earth echoes a line from Luke's gospel: "I tell you, if these [disciples] were silent, the stones would shout out." See Luke 19:40. Luke is writing about the praise of God at the moment of Jesus's entry into Jerusalem, but the potential cry of the stones could be interpreted to possess a more somber edge when read in the context of the Anthropocene. If we refuse to speak out about the ecological trauma of the Anthropocene, then it will be the stones that do the talking for us.

110. Skrimshire, "Anthropocene Fever," 153.

111. Narine, *Eco-Trauma Cinema*, 13.

112. Skrimshire, "Anthropocene Fever," 152.

113. Skrimshire, "Anthropocene Fever," 153.

114. Rambo, *Resurrecting Wounds*, 151.

Conclusion

1. James Hatley, *Suffering Witness: The Quandary of Responsibility After the Irreparable* (Albany: State University of New York Press, 2000), 4.

2. Shelly Rambo, *Spirit and Trauma: A Theology of Remaining* (Louisville, KY: Westminster John Knox Press, 2010), 143.

3. Paul Kingsnorth and Dougald Hine, "Uncivilisation: The Dark Mountain Manifesto," *The Dark Mountain Project* (2009), accessed 7 June 2024, https://dark-mountain.net/about/manifesto/.

4. Jem Bendell, "Deep Adaptation: A Map for Navigating Climate Tragedy," in *Deep Adaptation: Navigating the Realities of Climate Chaos*, ed. Jem Bendell and Rupert Read (Cambridge: Polity Press, 2021), 42–86, 72.

5. Bendell, "Deep Adaptation," 43.

6. Bendell suggests that relinquishment entails "people and communities letting go of certain assets, behaviours and beliefs where retaining them could make matters worse," including moving away from coastlines and giving up on certain expectations about what patterns of consumption are sustainable. Meanwhile, his proposal for resilience may well require "creative reinterpretation of identity and priorities." Just as with trauma, there is no return to how life was before. See Bendell, "Deep Adaptation," 72–73.

7. The predictions made by these movements may not be correct, but any narrative of collapse is likely to be culturally and politically taboo regardless of whether or not it turns out to be true. See Susanne C. Moser, "The Work After 'It's Too Late' (to Prevent Dangerous Climate Change)," *Wiley Interdisciplinary Reviews: Climate Change* 11 (2020): 1–11, 3–4.

8. The question of whether it is "too late" to prevent dangerous climate change, and the potential for ensuing societal collapse, is being debated in serious scientific circles, but there is no straightforward answer. See Moser, "The Work After 'It's Too Late,'" 1.

9. Shelly Rambo, *Resurrecting Wounds: Living in the Afterlife of Trauma* (Waco, TX: Baylor University Press, 2017), 145.

10. Rachel Becker, "Why Scare Tactics Won't Stop Climate Change," *The Verge* (2017), accessed 16 September 2024, https://www.theverge.com/2017/7/11/15954106/doomsday-climate-science-apocalypse-new-york-magazine-response.

11. Saffron J. O'Neill and Sophie Nicholson-Cole, "'Fear Won't Do It': Promoting Positive Engagement with Climate Change Through Visual and Iconic Representations," *Science Communication* 30 (2009): 355–79; Susie Wang et al., "Public Engagement with Climate Imagery in a Changing Digital Landscape," *Wiley Interdisciplinary Reviews: Climate Change* 9 (2018): 1–18.

12. Moser, "The Work After 'It's Too Late,'" 5.

13. Rebecca Solnit, *Hope in the Dark: Untold Histories, Wild Possibilities* (Edinburgh: Canongate Books, 2016), 4–5.

14. Blanche Verlie, *Learning to Live with Climate Change: From Anxiety to Transformation* (London: Routledge, 2022), 114.

15. Holmes Rolston III, *Science & Religion: A Critical Survey* (Philadelphia: Templeton Foundation Press, 2006), 144–46, 289–93; Holmes Rolston III, "Redeeming a Cruciform Nature," *Zygon* 53 (2018): 739–51.

16. Celia Deane-Drummond, *Christ and Evolution: Wonder and Wisdom* (London: SCM Press, 2009), 172.

17. Deanna A. Thompson, *Glimpsing Resurrection: Cancer, Trauma, and Ministry* (Louisville, KY: Westminster John Knox Press, 2018), 147.

18. Thompson, *Glimpsing Resurrection*, 147; Miguel A. De La Torre, *Embracing Hopelessness* (Minneapolis: Fortress Press, 2017), 141.

19. De La Torre, *Embracing Hopelessness*, 141.

20. De La Torre, *Embracing Hopelessness*, 139.

21. Rambo, *Resurrecting Wounds*, 8; Serene Jones, *Trauma and Grace: Theology in a Ruptured World*, 2nd Edition (Louisville, KY: Westminster John Knox Press, 2009), 137.

22. De La Torre, *Embracing Hopelessness*, 4.

23. De La Torre, *Embracing Hopelessness*, 96.

24. Donna J. Haraway, *Staying with the Trouble: Making Kin in the Chthulucene* (Durham, NC: Duke University Press, 2016), 1.

25. Catherine Keller, *Political Theology of the Earth: Our Planetary Emergency and the Struggle for a New Public* (New York: Columbia University Press, 2018), 90.

26. Panu Pihkala, "The Cost of Bearing Witness to the Environmental Crisis: Vicarious Traumatization and Dealing with Secondary Traumatic Stress Among Environmental Researchers," *Social Epistemology* 34 (2020): 86–100, 94. See also:

Terry Eagleton, *Hope Without Optimism* (New Haven, CT: Yale University Press, 2015).

27. Solnit, *Hope in the Dark*, xii.

28. Solnit, *Hope in the Dark*, xi–xii.

29. Panu Pihkala, "Eco-Anxiety, Tragedy, and Hope: Psychological and Spiritual Dimensions of Climate Change," *Zygon* 53 (2018): 545–69, 555; Pamela R. McCarroll, *Waiting at the Foot of the Cross: Toward a Theology of Hope for Today* (Eugene, OR: Wipf & Stock, 2014), 5. Rebecca Solnit quotes from Virginia Woolf: "The future is dark, which is on the whole, the best thing the future can be, I think." Hope is "in the dark" not because it emerges from something terrible but because it remains inscrutable, dependent on an unknown and unknowable future. See Solnit, *Hope in the Dark*, 1.

30. See Keller, *Political Theology of the Earth*, 29, 90.

31. Jonathan Lear, *Radical Hope: Ethics in the Face of Cultural Devastation* (Cambridge, MA: Harvard University Press, 2006).

32. Lear, *Radical Hope*, 93. Rambo draws here on resources from beyond the Christian tradition; these are concerns that cut across cultural and religious affiliations.

33. Rambo, *Spirit and Trauma*, 168.

34. In fact, Lear is explicit about the relevance of Plenty Coups for the contemporary situation, writing that "the aim is to establish what *we* might legitimately hope at a time when the sense of purpose and meaning that has been bequeathed to us by our culture has collapsed." See Lear, *Radical Hope*, 104.

35. See Romans 4:18 on "hope against hope."

36. Pihkala, "The Cost of Bearing Witness," 94.

37. Stefan Skrimshire, "Anthropocene Fever: Memory and the Planetary Archive," in *Religion and the Anthropocene*, ed. Celia Deane-Drummond et al. (Eugene, OR: Cascade Books, 2017), 138–54, 153.

38. Rambo, *Resurrecting Wounds*, 151.

39. Rambo, *Spirit and Trauma*, 170. Similarly, Dirk G. Lange maintains that liturgical repetition of the Christ event constitutes a "thrust toward responsibility." See Lange, *Trauma Recalled: Liturgy, Disruption, and Theology* (Minneapolis: Fortress Press, 2010), 98. Lange even goes one step further, suggesting that the only response to trauma is an ethical response given that its inaccessibility results in a failure to adequately represent it. See Lange, *Trauma Recalled*, 105.

40. Rambo, *Resurrecting Wounds*, 151.

41. Verlie, *Learning to Live with Climate Change*, 74.

42. Kelly Oliver, *Witnessing: Beyond Recognition* (Minneapolis: University of Minnesota Press, 2001), 91.

43. Michael Richardson, "Climate Trauma, or the Affects of the Catastrophe to Come," *Environmental Humanities* 10 (2018): 1–19, 4. As James Hatley puts it, "by witness is meant a mode of responding to the other's plight that exceeds

epistemological determination and becomes an ethical involvement." See Hatley, *Suffering Witness*, 3.

44. See Richardson, "Climate Trauma," 149.

45. Joanna Zylinska's minimal ethics seeks "to keep philosophizing as if against all odds, to look for signs of life in the middle of an apocalypse." She continues, "minimal ethics for the Anthropocene is to serve as a caution against understanding the Anthropocene too well and too quickly, and against knowing precisely how to solve the problems it represents." See Zylinska, *Minimal Ethics for the Anthropocene* (Ann Arbor, MI: Open Humanities Press, 2014), 13, 124.

46. Verlie, *Learning to Live with Climate Change*, 116.

47. Hatley, *Suffering Witness*, 3.

48. Rambo, *Spirit and Trauma*, 37–44.

49. Verlie, *Learning to Live with Climate Change*, 80. Much of the work of climate and conservation scientists could be construed as witnessing in this vein. They are not just discovering facts about the natural world but witnessing to traumas that have previously gone unrecognized. There are similarities here with Bruno Latour's suggestion that different elements of the natural world deserve representation (or witness) in a "Parliament of Things." The role of an oceanographer, for example, is to bear witness to the oceans in this forum. See Latour, *Facing Gaia: Eight Lectures on the New Climatic Regime*, trans. Catherine Porter (Cambridge: Polity Press, 2017), 255–66.

50. Stacy Alaimo, *Undomesticated Ground: Recasting Nature as Feminist Space* (Ithaca, NY: Cornell University Press, 2000), 182, quoted in Celia Deane-Drummond, *Eco-Theology* (London: Darton, Longman and Todd, 2008), 149.

51. Deane-Drummond, *Eco-Theology*, 89.

52. Dori Laub, "Bearing Witness or the Vicissitudes of Listening," in *Testimony: Crises of Witnessing in Literature, Psychoanalysis, and History*, ed. Shoshana Felman and Dori Laub (Abingdon: Taylor & Francis, 1992), 57–74, 57.

53. Karen O'Donnell, *Broken Bodies: The Eucharist, Mary, and the Body in Trauma Theology* (London: SCM Press, 2018), 3. For Dirk Lange, the practice of Christian liturgy does not master or control the trauma of the Christ event, but rather disrupts and disseminates an event that cannot be systematized. As he writes, "repetition (liturgical repetition) does not repeat or capture the Christ event (life, death, and resurrection) but, paradoxically, repeats or iterates the impossibility of repeating that event." See Lange, *Trauma Recalled*, 94.

54. O'Donnell, *Broken Bodies*, 191.

55. Catherine Pickstock, *Repetition and Identity*, Literary Agenda (Oxford: Oxford University Press, 2013), 177.

56. St. Columban's Mission Society, "Stations of the Forests" (St. Columban's Mission Society, 2011), 1–28, accessed 9 March 2021, https://www.columban.org.au/assets/files/cmi/columban stations of the forest booklet.pdf. For further discussion of "Stations of the Forests" see: Timothy A. Middleton, "Christic Witnessing: A Practical Response to Ecological Trauma," *Practical Theology* 15 (2022): 420–31.

57. Cymene Howe and Dominic Boyer, "Death of a Glacier," *Anthropology News* 61 (2020): 16–21.

58. O'Donnell, *Broken Bodies*, 25, 191.

59. O'Donnell, *Broken Bodies*, 167–69; Lange, *Trauma Recalled*, 149–50. Although Lange also notes that "the silence of this text [the *Didache*] about the cross witnesses to the trauma of such an execution." See Lange, *Trauma Recalled*, 151.

60. Deane-Drummond, *Eco-Theology*, 60. However, this eucharistic cosmology also needs to be mindful of the concern I mentioned in the Introduction about cosmic Christology constituting a colonial act of imposition.

61. Elizabeth Theokritoff, "Creation and Salvation in Orthodox Worship," *Ecotheology* 10 (2001): 97–108, 105. Similarly, John Chryssavgis recounts how, in the context of ecological crisis, "it is now clear that grave 'fissures' and 'faults' have appeared on the face of the earth." Chryssavgis emphasizes the place of sacramental theology in "restoring the shattered image of the world." Yet perhaps the broken host can also serve as a witness to these fissures and faults, without rushing too quickly to restoration. See Chryssavgis, *Creation as Sacrament: Reflections on Ecology and Spirituality* (New York: T & T Clark, 2019), 1.

62. Pierre Teilhard de Chardin, *Hymn of the Universe* (New York: Harper & Row, 1965), 20.

63. O'Donnell, *Broken Bodies*, 152.

64. Karen O'Donnell, "The Voices of the Marys: Towards a Method in Feminist Trauma Theologies," in *Feminist Trauma Theologies: Body, Scripture & Church in Critical Perspective*, ed. Karen O'Donnell and Katie Cross (London: SCM Press, 2020), 3–20, 16.

Bibliography

Abbott, Roger, and Bob White. "Haiti—An Unnatural Disaster." *Ethics in Brief* 18 (2013): 1–4
Airenti, Gabriella. "The Development of Anthropomorphism in Interaction: Intersubjectivity, Imagination, and Theory of Mind." *Frontiers in Psychology* 9 (2018): 1–13. https://doi.org/10.3389/fpsyg.2018.02136.
Alaimo, Stacy. *Undomesticated Ground: Recasting Nature as Feminist Space*. Ithaca, NY: Cornell University Press, 2000.
Alexander, Jeffrey C. "Toward a Theory of Cultural Trauma." In *Cultural Trauma and Collective Identity*, edited by Jeffrey C. Alexander, Ron Eyerman, Bernhard Giesen, Neil J. Smesler, and Piotr Sztompka, 1–30. Berkeley: University of California Press, 2004.
Andermahr, Sonya. "'Decolonizing Trauma Studies: Trauma and Postcolonialism'—Introduction." *Humanities* 4 (2015): 500–505. https://doi.org/10.3390/h40 40500.
Anthropocene Working Group. "What Is the Anthropocene? – Current Definition and Status." *Subcommission on Quaternary Stratigraphy*. Accessed 12 January 2022. https://quaternary.stratigraphy.org/working-groups/anthropocene.
Archer, David. "Fate of Fossil Fuel CO2 in Geologic Time." *Journal of Geophysical Research C: Oceans* 110 (2005): 1–6. https://doi.org/10.1029/2004JC002625.
Arendt, Hannah. *Responsibility and Judgment*. Edited by Jerome Kohn. New York: Schocken Books, 2003.
Ashworth, Justin. "How Divine Solidarity Liberates." *Scottish Journal of Theology* 72 (2019): 324–34. https://doi.org/10.1017/S003693061900036X.
Augustine. *Confessions*. Oxford World's Classics. Translated by Henry Chadwick. Oxford: Oxford University Press, 1998.
Baldwin, Jennifer. *Trauma-Sensitive Theology: Thinking Theologically in the Era of Trauma*. Eugene, OR: Wipf & Stock, 2018.

Balthasar, Hans Urs von. *First Glance at Adrienne von Speyr*. Translated by Antje Lawry and Sergia Englund. San Francisco: Ignatius Press, 1981.

Balthasar, Hans Urs von. *Heart of the World*. Translated by Erasmo S. Leiva. San Francisco: Ignatius Press, 1979.

Balthasar, Hans Urs von. *The Moment of Christian Witness*. Communio Books. Translated by Richard Beckley. San Francisco: Ignatius Press, 1994.

Balthasar, Hans Urs von. *Spirit and Institution, Volume 4 of Explorations in Theology*. Twentieth Century Religious Thought. Translated by Edward T. Oakes. San Francisco: Ignatius Press, 1995.

Bauckham, Richard. "The Incarnation and the Cosmic Christ." In *Incarnation: On the Scope and Depth of Christology*, edited by Niels Henrik Gregersen, 25–58. Minneapolis: Fortress Press, 2015.

Bauckham, Richard. *Jesus and the Eyewitnesses: The Gospels as Eyewitness Testimony*, 2nd Edition. Grand Rapids, MI: Eerdmans, 2017.

Bauckham, Richard. "Joining Creation's Praise of God." *Ecotheology* 7 (2002): 45–59. https://doi.org/10.1558/ecotheology.v7i1.45.

Bauckham, Richard. "The Story of the Earth According to Paul: Romans 8:18–23." *Review & Expositor* 108 (2011): 91–97. https://doi.org/10.1177/003463731110800109.

Bauman, Whitney A. "Sourdough Time and the Time of Protest: Reflections on the Pace of Planetary Becoming." *Minding Nature* 13 (2020): 94–99.

Bechtel, Trevor, Matthew Eaton, and Timothy Harvie. "Introduction." In *Encountering Earth: Thinking Theologically With a More-Than-Human World*, edited by Trevor Bechtel, Matthew Eaton, and Timothy Harvie, 1–13. Eugene, OR: Cascade Books, 2018.

Beck, Shawn Sanford. *Christian Animism*. Winchester: Christian Alternative, 2015.

Becker, Rachel. "Why Scare Tactics Won't Stop Climate Change." *The Verge*, 2017. Accessed 16 September 2024. https://www.theverge.com/2017/7/11/15954106/doomsday-climate-science-apocalypse-new-york-magazine-response.

Behr, John. *Irenaeus of Lyons: Identifying Christianity*. Oxford: Oxford University Press, 2013.

Bekoff, Marc. "Animal Emotions: Exploring Passionate Natures." *BioScience* 50 (2000): 861–70. https://doi.org/10.1641/0006-3568(2000)050[0861:AEEPN]2.0.CO;2.

Bendell, Jem. "Deep Adaptation: A Map for Navigating Climate Tragedy." In *Deep Adaptation: Navigating the Realities of Climate Chaos*, edited by Jem Bendell and Rupert Read, 42–86. Cambridge: Polity Press, 2021.

Bennett, Jane. *Vibrant Matter: A Political Ecology of Things*. Durham, NC: Duke University Press, 2010.

Berry, Helen L., Thomas D. Waite, Keith B. G. Dear, Anthony G. Capon, and Virginia Murray, "The Case for Systems Thinking About Climate Change and Mental Health." *Nature Climate Change* 8 (2018): 282–90. https://doi.org/10.1038/s41558-018-0102-4.

Beste, Jennifer Erin. *God and the Victim: Traumatic Intrusions on Grace and Freedom*. Oxford: Oxford University Press, 2007.
Betcher, Sharon V. "Becoming Flesh of My Flesh: Feminist and Disability Theologies on the Edge of Posthumanist Discourse." *Journal of Feminist Studies in Religion* 26 (2010): 107–18. https://doi.org/10.2979/fsr.2010.26.2.107.
Betcher, Sharon V. *Spirit and the Politics of Disablement*. Minneapolis: Fortress Press, 2007.
Bjornerud, Marcia. *Timefulness: How Thinking Like a Geologist Can Help Save the World*. Princeton, NJ: Princeton University Press, 2018.
Boase, Elizabeth, and Christopher G. Frechette, eds. *Bible Through the Lens of Trauma*. Atlanta: SBL Press, 2016.
Boase, Elizabeth, and Christopher G. Frechette. "Defining 'Trauma' as a Useful Lens for Biblical Interpretation." In *Bible through the Lens of Trauma*, edited by Elizabeth Boase and Christopher G. Frechette, 1–24. Atlanta: SBL Press, 2016.
Boer, Matthias M., Victor Resco de Dios, and Ross A. Bradstock. "Unprecedented Burn Area of Australian Mega Forest Fires." *Nature Climate Change* 10 (2020): 170–72. https://doi.org/10.1038/s41558-020-0710-7.
Boff, Leonardo. *Cry of the Earth, Cry of the Poor*. Ecology and Justice Series. Maryknoll, NY: Orbis Books, 1997.
Bonaventure. *The Sunday Sermons of St. Bonaventure*. Bonaventure Texts in Translation Series. Edited by Robert F. Karris. Translated by Timothy J. Johnson. St. Bonaventure, NY: Franciscan Institute, 2008.
Boynton, Eric. "Evil, Trauma, and the Building of Absences." In *Trauma and Transcendence: Suffering and the Limits of Theory*, 1st Edition, edited by Eric Boynton and Peter Capretto, 83–101. New York: Fordham University Press, 2018. https://doi.org/10.2307/j.ctv19x52c.7.
Boynton, Eric, and Peter Capretto. "Introduction." In *Trauma and Transcendence: Suffering and the Limits of Theory*, 1st Edition, edited by Eric Boynton and Peter Capretto, 1–14. New York: Fordham University Press, 2018. https://doi.org/10.2307/j.ctv19x52c.3.
Braaten, Laurie J. "Earth Community in Hosea 2." In *The Earth Story in the Psalms and the Prophets*, edited by Norman C. Habel, 185–203. Sheffield: Sheffield Academic Press, 2001.
Bradshaw, Gay. *Elephants on the Edge: What Animals Teach Us About Humanity*. New Haven, CT: Yale University Press, 2009.
Bridgers, Lynn. "The Resurrected Life: Roman Catholic Resources in Posttraumatic Pastoral Care." *International Journal of Practical Theology* 15 (2011): 38–56. https://doi.org/10.1515/IJPT.2011.025.
Brulle, Robert J., and Kari Marie Norgaard. "Avoiding Cultural Trauma: Climate Change and Social Inertia." *Environmental Politics* (2019): 1–23. https://doi.org/10.1080/09644016.2018.1562138.
Buber, Martin. *I and Thou*. Translated by Ronald Gregor Smith. Edinburgh: T & T Clark, [1923] 2004.

Burghardt, Gordon M. "Animal Awareness: Current Perceptions and Historical Perspective." *American Psychologist* 40 (1985): 905–19 https://doi.org/10.1037/0003-066X.40.8.905.

Cadwallader, Alan. "'And the Earth Shook'—Mortality and Ecological Diversity: Interpreting Jesus' Death in Matthew's Gospel." In *Biodiversity and Ecology as Interdisciplinary Challenge*, edited by Denis Edwards and Mark William Worthing, 45–54. Adelaide: ATF Press, 2004.

Calvin, John. *Commentary on the Gospel According to John, Volume 2*. Translated by William Pringle. Grand Rapids, MI: Eerdmans, 1956.

Cannon, Katie G. "Womanist Perspectival Discourse and Cannon Formation." *Journal of Feminist Studies in Religion* 9 (1993): 29–37. https://www.jstor.org/stable/25002198.

Carley, Keith. "Ezekiel's Formula of Desolation: Harsh Justice for the Land/Earth." In *The Earth Story in the Psalms and the Prophets*, edited by Norman C. Habel, 143–57. Sheffield: Sheffield Academic Press, 2001.

Carr, David McLain. *Holy Resilience: The Bible's Traumatic Origins*. New Haven, CT: Yale University Press, 2014.

Carrington, Damian. "Why the Guardian Is Changing the Language It Uses About the Environment." *The Guardian*, 2019. Accessed 11 April 2024. https://www.theguardian.com/environment/2019/may/17/why-the-guardian-is-changing-the-language-it-uses-about-the-environment.

Caruth, Cathy. "Recapturing the Past: Introduction," In *Trauma: Explorations in Memory*, edited by Cathy Caruth, 151–57. Baltimore, MD: Johns Hopkins University Press, 1995.

Caruth, Cathy. "Trauma and Experience: Introduction." In *Trauma: Explorations in Memory*, edited by Cathy Caruth, 3–12. Baltimore, MD: Johns Hopkins University Press, 1995.

Carvalhaes, Cláudio. "Colonization, Trauma and Prayers: Towards a Collective Healing." In *Bearing Witness: Intersectional Perspectives on Trauma Theology*, edited by Karen O'Donnell and Katie Cross, 294–310. London: SCM Press, 2022.

Castelli, Elizabeth. *Imitating Paul: A Discourse of Power*. Literary Currents in Biblical Interpretation. Louisville, KY: Westminster John Knox Press, 1991.

Castelli, Elizabeth. *Martyrdom and Memory: Early Christian Culture Making*. Gender, Theory, and Religion. New York: Columbia University Press, 2004.

Chakrabarty, Dipesh. "Humanities in the Anthropocene: The Crisis of an Enduring Kantian Fable." *New Literary History* 47 (2016): 377–97. https://doi.org/10.1353/nlh.2016.0019.

Chopp, Rebecca S. "Theology and the Poetics of Testimony." In *Converging on Culture: Theologians in Dialogue with Cultural Analysis and Criticism*, edited by Delwin Brown, Sheila Greeve Davaney, and Kathryn Tanner, 56–70. Oxford: Oxford University Press, 2001.

Chryssavgis, John. *Cosmic Grace, Humble Prayer: The Ecological Vision of the Green Patriarch*. Grand Rapids, MI: Eerdmans, 2003.

Chryssavgis, John. *Creation as Sacrament: Reflections on Ecology and Spirituality*. New York: T & T Clark, 2019.
Clayton, Susan, Christie Manning, Kirra Krygsman, and Meighen Speiser. *Mental Health and Our Changing Climate: Impacts, Implications, and Guidance*. Washington, D.C.: American Psychological Association and ecoAmerica, 2017.
Clayton, Susan, Christie Manning, Meighen Speiser, and Alison Nicole Hill. *Mental Health and Our Changing Climate: Impacts, Inequities, Responses*. Washington, D.C.: American Psychological Association and ecoAmerica, 2021.
Cole-Turner, Ronald. "Incarnation Deep and Wide: A Response to Niels Gregersen." *Theology and Science* 11 (2013): 424–35. https://doi.org/10.1080/14746700.2013.836886.
Collicutt McGrath, Joanna. "Post-Traumatic Growth and the Origins of Early Christianity." *Mental Health, Religion and Culture* 9 (2006): 291–306. https://doi.org/10.1080/13694670600615532.
Collins, Natalie. "Broken or Superpowered? Traumatized People, Toxic Doublethink and the Healing Potential of Evangelical Christian Communities." In *Feminist Trauma Theologies: Body, Scripture & Church in Critical Perspective*, edited by Karen O'Donnell and Katie Cross, 195–221. London: SCM Press, 2020.
Cone, James H. *The Cross and the Lynching Tree*. Maryknoll, NY: Orbis Books, 2011.
Cone, James H. "Whose Earth Is It, Anyway?" *Cross Currents* 50 (2000): 36–46. https://doi.org/10.4324/9781315625546.
Conradie, Ernst M. "Resurrection, Finitude, and Ecology." In *Resurrection: Theological and Scientific Assessments*, edited by Ted Peters, Robert John Russell, and Michael Welker, 277–96. Grand Rapids, MI: Eerdmans, 2002.
Copeland, Rebecca L. "'Their Leaves Shall Be for Healing': Ecological Trauma and Recovery in Ezekiel 47:1–12." *Biblical Theology Bulletin* 49 (2019): 214–22. https://doi.org/10.1177/0146107919877639.
Craps, Stef. "Climate Trauma." In *The Routledge Companion to Literature and Trauma*, edited by Colin Davis and Hanna Meretoja, 275–84. London: Routledge, 2020.
Craps, Stef. *Postcolonial Witnessing: Trauma Out of Bounds*. Basingstoke: Palgrave Macmillan, 2013.
Creamer, Deborah Beth. "Including All Bodies in the Body of God: Disability and the Theology of Sallie McFague." *Journal of Religion, Disability & Health* 9 (2006): 55–69. https://doi.org/10.1300/J095v09n04_04.
Cross, Katie, and Karen O'Donnell. "Introduction." In *Bearing Witness: Intersectional Perspectives on Trauma Theology*, edited by Karen O'Donnell and Katie Cross, 1–9. London: SCM Press, 2022.
Crutzen, Paul J. "Geology of Mankind." *Nature* 415 (2002): 23. https://doi.org/10.1038/415023a.
Crutzen, Paul J., and Eugene F. Stoermer. "The 'Anthropocene.'" *International Geosphere-Biosphere Programme Global Change Newsletter* (2000), 17–18. http://

www.igbp.net/download/18.316f18321323470177580001401/1376383088452/NL41.pdf.

Cullmann, Oscar. *Christ and Time: The Primitive Christian Conception of Time and History*. Translated by Floyd V. Filson. London: SCM Press, 1951.

Cunningham, David S. "The Way of All Flesh: Rethinking the Imago Dei." In *Creaturely Theology: On God, Humans and Other Animals*, edited by Celia Deane-Drummond and David Clough, 100–117. London: SCM Press, 2009.

Daly, Mary. *Beyond God the Father: Towards a Philosophy of Women's Liberation*. Wellingborough: The Women's Press, 1986.

Daly-Denton, Margaret. *John, an Earth Bible Commentary: Supposing Him to Be the Gardener*. London: Bloomsbury Academic, 2017.

De La Torre, Miguel A. *Embracing Hopelessness*. Minneapolis: Fortress Press, 2017.

Deane-Drummond, Celia. *Christ and Evolution: Wonder and Wisdom*. London: SCM Press, 2009.

Deane-Drummond, Celia. "Deep Incarnation Between Balthasar and Bulgakov: The Form of Beauty and the Wisdom of God." In *Envisioning the Cosmic Body of Christ: Embodiment, Plurality and Incarnation*, edited by Aurica Jax and Saskia Wendel, 101–13. Abingdon: Routledge, 2019. https://doi.org/10.4324/9780429340604-9.

Deane-Drummond, Celia. *Eco-Theology*. London: Darton, Longman and Todd, 2008.

Deane-Drummond, Celia. "Performing the Beginning in the End: A Theological Anthropology for the Anthropocene." In *Religion and the Anthropocene*, edited by Celia Deane-Drummond, Sigurd Bergmann, and Markus Vogt, 173–87. Eugene, OR: Cascade Books, 2017.

Deane-Drummond, Celia. *A Primer in Ecotheology: Theology for a Fragile Earth*. Eugene, OR: Cascade Books, 2017.

Deane-Drummond, Celia. "Who on Earth Is Jesus Christ? Plumbing the Depths of Deep Incarnation." In *Christian Faith and the Earth: Current Paths and Emerging Horizons in Ecotheology*, edited by Ernst M. Conradie, Sigurd Bergmann, Celia Deane-Drummond, and Denis Edwards, 31–50. New York: Bloomsbury T & T Clark, 2014. https://doi.org/10.5040/9780567659613.ch-003.

Deane-Drummond, Celia. *The Wisdom of the Liminal: Evolution and Other Animals in Human Becoming*. Grand Rapids, MI: Eerdmans, 2014.

Deane-Drummond, Celia. "The Wisdom of Fools? A Theo-Dramatic Interpretation of Deep Incarnation." In *Incarnation: On the Scope and Depth of Christology*, edited by Niels Henrik Gregersen, 177–202. Minneapolis: Fortress Press, 2015.

DeGruy, Joy. *Post Traumatic Slave Syndrome: America's Legacy of Enduring Injury and Healing*. Portland, OR: Joy DeGruy Publications, 2005.

Deloria, Vine, Jr. "If You Think About It, You Will See That It Is True." In *Spirit & Reason: The Vine Deloria, Jr., Reader*. Edited by Barbara Deloria, Kristen Foehner, and Samuel Scinta. Golden, CO: Fulcrum, 1999.

Descola, Philippe. *Beyond Nature and Culture*. Translated by Janet Lloyd. Chicago: University of Chicago Press, 2013.
Deslauriers, Jessica, Mate Toth, Andre Der-Avakian, and Victoria B. Risbrough. "Current Status of Animal Models of Posttraumatic Stress Disorder: Behavioral and Biological Phenotypes, and Future Challenges in Improving Translation." *Biological Psychiatry* 83 (2018): 895–907. https://doi.org/10.1016/j.biopsych.2017.11.019.
Dodd, C. H. *The Parables of the Kingdom*, Revised Edition. London: Nisbet, 1961.
Doehring, Carrie. *Internal Desecration: Traumatization and Representation of God*. Lanham, MD: University Press of America, 1993.
Douglas, Kelly Brown. "Foreword." In *Trauma and Grace: Theology in a Ruptured World*, 2nd Edition, by Serene Jones, vii–x. Louisville, KY: Westminster John Knox Press, 2019.
Eagleton, Terry. *Hope Without Optimism*. New Haven, CT: Yale University Press, 2015. https://doi.org/10.12987/9780300220254.
Earth Bible Team. "The Voice of Earth: More than Metaphor?" In *The Earth Story in the Psalms and the Prophets*, edited by Norman C. Habel, 23–28. Sheffield: Sheffield Academic Press, 2001.
Eaton, Heather. "An Earth-Centric Theological Framing for Planetary Solidarity." In *Planetary Solidarity: Global Women's Voices on Christian Doctrine and Climate Justice*, edited by Grace Ji-Sun Kim and Hilda P. Koster, 19–44. Minneapolis: Fortress Press, 2017.
Eaton, Matthew. "Conclusion: Ecocide as Deicide: Eschatological Lamentation and the Possibility of Hope." In *Integral Ecology for a Sustainable World: Dialogues with Laudato Si'*, edited by Dennis O'Hara, Matthew Eaton, and Michael Ross, 359–72. Lanham, MD: Lexington Books, 2019.
Eaton, Matthew. "Enfleshed in Cosmos and Earth: Re-Imagining the Depth of the Incarnation." *Worldviews: Global Religions, Culture, and Ecology* 18 (2014): 230–54. https://doi.org/10.1163/15685357-01803002.
Eaton, Matthew. "Enfleshing Cosmos and Earth: An Ecological Theology of Divine Incarnation." Unpublished doctoral thesis (University of St. Michael's College, 2017). Accessed 6 November 2018. https://tspace.library.utoronto.ca/bitstream/1807/81412/6/Eaton_Matthew_201711_PhD_thesis.pdf.
Edmondson, Stephen. *Calvin's Christology*. Cambridge: Cambridge University Press, 2004.
Edwards, Denis. *Deep Incarnation: God's Redemptive Suffering with Creatures*. Maryknoll, NY: Orbis Books, 2019.
Edwards, Denis. *Ecology at the Heart of Faith*. Maryknoll, NY: Orbis Books, 2006.
Edwards, Denis. "Every Sparrow That Falls to the Ground: The Cost of Evolution and the Christ-Event." *Ecotheology* 11 (2006): 103–23. https://doi.org/10.1558/ecot.2006.11.1.103.
Erikson, Kai. *A New Species of Trouble: Explorations in Disaster, Trauma, and Community*. New York: W. W. Norton, 1994.

Erikson, Kai. *Everything in Its Path: Destruction of Community in the Buffalo Creek Flood*. New York: Simon & Schuster, 1976.

Erikson, Kai. "Notes on Trauma and Community." In *Trauma: Explorations in Memory*, edited by Cathy Caruth, 183–99. Baltimore, MD: Johns Hopkins University Press, 1995.

Extinction Rebellion. *This Is Not a Drill: The Extinction Rebellion Handbook*. London: Penguin Books, 2019.

Eyerman, Ron. *Is This America? Katrina as Cultural Trauma*. The Katrina Bookshelf. Austin: University of Texas Press, 2015. https://doi.org/10.7560/303689.

Farley, Wendy. *Tragic Vision and Divine Compassion: A Contemporary Theodicy*. Louisville, KY: Westminster John Knox Press, 1990.

Fertel, Randy. "Hearing the Bugle's Call: Hurricane Katrina, the BP Oil Spill, and the Effects of Trauma." In *Environmental Disasters and Collective Trauma*, edited by Nancy Cater and Stephen Foster, 91–115. New Orleans: Spring Journal, 2012.

Feuerbach, Ludwig. *The Essence of Christianity*. Translated by George Eliot. Mineola, NY: Dover Publications, [1881] 2008.

Fiddes, Paul S. *Participating in God: A Pastoral Doctrine of the Trinity*. London: Darton, Longman & Todd, 2000.

Fisher, Andy. *Radical Ecopsychology: Psychology in the Service of Life*, 2nd Edition. Albany: State University of New York Press, 2013.

Foster, John. *After Sustainability: Denial, Hope, Retrieval*. Abingdon: Routledge, 2015.

Fredengren, Christina. "Re-Wilding the Environmental Humanities: A Deep Time Comment." *Current Swedish Archaeology* 26 (2018): 50–60.

Fretheim, Terence E. *God and World in the Old Testament: A Relational Theology of Creation*. Nashville, TN: Abingdon Press, 2005.

Fuchs, Thomas. *Ecology of the Brain: The Phenomenology and Biology of the Embodied Mind*. Oxford: Oxford University Press, 2018.

Galea, Sandro, Chris R. Brewin, Michael Gruber et al. "Exposure to Hurricane-Related Stressors and Mental Illness after Hurricane Katrina." *Archives of General Psychiatry* 64 (2007): 1427–34. https://doi.org/10.1001/archpsyc.64.12.1427.

Ganzevoort, R. Ruard, and Srdjan Sremac, eds. *Trauma and Lived Religion: Transcending the Ordinary*. Cham, Switzerland: Palgrave Macmillan, 2019.

Ginn, Franklin, Michelle Bastian, David Farrier, and Jeremy Kidwell. "Introduction: Unexpected Encounters with Deep Time." *Environmental Humanities* 10 (2018): 213–25. https://doi.org/10.1215/22011919-4385534.

Gottlieb, Roger S. *Morality and the Environmental Crisis*. Cambridge: Cambridge University Press, 2019.

Gould, Stephen Jay. *Rocks of Ages: Science and Religion in the Fullness of Life*. London: Vintage, 2002.

Graham, Elaine. "After the Fire, the Voice of God: Speaking of God After Tragedy and Trauma." In *Tragedies and Christian Congregations: The Practical Theology of Trauma*, edited by Megan Warner, Christopher Southgate, Carla A. Grosch-Miller, and Hilary Ison, 13–27. Abingdon: Routledge, 2019.

Gregersen, Niels Henrik. "Christology." In *Climate Change and Systematic Theology: Ecumenical Perspectives*, edited by Michael Northcott and Peter Scott. 33–50. New York: Routledge, 2014.

Gregersen, Niels Henrik. "The Cross of Christ in an Evolutionary World." *Dialog: A Journal of Theology* 40 (2001): 192–207. https://doi.org/10.1111/0012-2033.00075.

Gregersen, Niels Henrik. "Cur Deus Caro: Jesus and the Cosmos Story." *Theology and Science* 11 (2013): 370–93. https://doi.org/10.1080/14746700.2013.836891.

Gregersen, Niels Henrik. "Deep Incarnation and Kenosis: In, With, Under, and As: A Response to Ted Peters." *Dialog* 52 (2013): 251–62. https://doi.org/10.1111/dial.12050.

Gregersen, Niels Henrik. "Deep Incarnation: From Deep History to Post-Axial Religion." *HTS Teologiese Studies / Theological Studies* 72 (2016): 1–12. https://doi.org/10.4102/hts.v72i4.3428.

Gregersen, Niels Henrik. "Deep Incarnation: The Logos Became Flesh." In *Transformative Theological Perspectives*, edited by Karen L. Bloomquist, 167–81. Minneapolis: Lutheran University Press, 2010.

Gregersen, Niels Henrik. "Deep Incarnation: Opportunities and Challenges." In *Incarnation: On the Scope and Depth of Christology*, edited by Niels Henrik Gregersen, 361–79. Minneapolis: Fortress Press, 2015.

Gregersen, Niels Henrik. "Deep Incarnation: Why Evolutionary Continuity Matters in Christology." *Toronto Journal of Theology* 26 (2010): 173–88. https://doi.org/10.3138/tjt.26.2.173.

Gregersen, Niels Henrik. "The Emotional Christ: Bonaventure and Deep Incarnation." *Dialog* 55 (2016): 247–61. https://doi.org/10.1111/dial.12261.

Gregersen, Niels Henrik. "The Extended Body: The Social Body of Jesus According to Luke." *Dialog* 51 (2012): 234–44. https://doi.org/10.1111/j.1540-6385.2012.00689.x.

Gregersen, Niels Henrik. "The Extended Body of Christ: Three Dimensions of Deep Incarnation." In *Incarnation: On the Scope and Depth of Christology*, edited by Niels Henrik Gregersen, 225–51. Minneapolis: Fortress Press, 2015.

Gregersen, Niels Henrik. "God, Matter, and Information: Towards a Stoicizing Logos Christology." In *Information and the Nature of Reality: From Physics to Metaphysics*, edited by Paul Davies and Niels Henrik Gregersen, 405–43. Cambridge: Cambridge University Press, 2010. https://doi.org/10.1017/CBO9781107589056.019.

Gregersen, Niels Henrik. "The Idea of Deep Incarnation: Biblical and Patristic Resources." In *To Discern Creation in a Scattering World*, edited by Frederiek Depoortere and Jacques Haers, 319–41. Leuven: Uitgeverij Peeters, 2013.

Gregersen, Niels Henrik, ed. *Incarnation: On the Scope and Depth of Christology*. Minneapolis: Fortress Press, 2015.

Gregersen, Niels Henrik. "Introduction." In *Incarnation: On the Scope and Depth of Christology*, edited by Niels Henrik Gregersen, 1–21. Minneapolis: Fortress Press, 2015.

Gregersen, Niels Henrik. "The Twofold Assumption: A Response to Cole-Turner, Moritz, Peters and Peterson." *Theology and Science* 11 (2013): 455–68. https://doi.org/10.1080/14746700.2013.866476.

Gregory the Great. *Forty Gospel Homilies*. Monastic Studies Series. Translated by Dom David Hurst. Piscataway, NJ: Gorgias Press, 2009.

Griffiths, Paul J. *Christian Flesh*. Stanford, CA: Stanford University Press, 2018.

Habel, Norman C. "The Crucified Land: Towards Our Reconciliation With the Earth." *Colloquium* 28 (1996): 3–19.

Habel, Norman C., with the Earth Bible Team. "Where Is the Voice of Earth in Wisdom Literature?" In *The Earth Story in Wisdom Traditions*, edited by Norman C. Habel and Shirley Wurst, 23–34. Sheffield: Sheffield Academic Press, 2001.

Hamilton, Clive. "The Anthropocene as Rupture." *Anthropocene Review* 3 (2016): 93–106. https://doi.org/10.1177/2053019616634741.

Hamilton, Clive. "Crimes Against Nature: The Banality of Ethics in the Anthropocene." *ABC Religion & Ethics* (2015). Accessed 6 May 2019. https://www.abc.net.au/religion/crimes-against-nature-the-banality-of-ethics-in-the-anthropocene/10098110.

Hamilton, Clive. *Defiant Earth: The Fate of Humans in the Anthropocene*. Cambridge: Polity Press, 2017.

Hamilton, Clive. "A Letter from Canberra: The Apocalyptic Fires in Australia Signal Another Future." *Sierra: The Magazine of the Sierra Club* (January 2020). https://www.sierraclub.org/sierra/letter-canberra.

Hamilton, Clive, Christophe Bonneuil, and François Gemenne. "Thinking the Anthropocene." In *The Anthropocene and the Global Environment Crisis: Rethinking Modernity in a New Epoch*, edited by Clive Hamilton, Christophe Bonneuil, and François Gemenne, 1–13. Abingdon: Routledge, 2015.

Hamilton, Clive, and Jacques Grinevald. "Was the Anthropocene Anticipated?" *Anthropocene Review* 2 (2015): 59–72. https://doi.org/10.1177/2053019614567155.

Haraway, Donna J. *Staying with the Trouble: Making Kin in the Chthulucene*. Durham, NC: Duke University Press, 2016.

Harding, Stephan. *Animate Earth: Science, Intuition and Gaia*. Totnes: Green Books, 2006.

Harding, Stephan. "Animate Earth." In *Earthy Realism: The Meaning of Gaia*, edited by Mary Midgley, 23–29. Exeter: Imprint Academic, 2007.

Harris, Melanie L. *Ecowomanism: African American Women and Earth-Honoring Faiths*. Ecology and Justice. Maryknoll, NY: Orbis Books, 2017.

Harvey, Graham. *Animism: Respecting the Living World*. New York: Columbia University Press, 2006.

Harvey, Mary R. "An Ecological View of Psychological Trauma and Trauma Recovery." *Journal of Traumatic Stress* 9 (1996): 3–23. https://doi.org/10.1002/jts.2490090103.

Harvey, Mary R. "Towards an Ecological Understanding of Resilience in Trauma Survivors." *Journal of Aggression, Maltreatment & Trauma* 14 (2007): 9–32. https://doi.org/10.1300/J146v14n01_02.

Hatley, James. *Suffering Witness: The Quandary of Responsibility After the Irreparable*. Albany: State University of New York Press, 2000.

Hayes, Katherine Murphey. *The Earth Mourns: Prophetic Metaphor and Oral Aesthetic*. Atlanta: Society of Biblical Literature, 2002.

Hayhoe, Katharine. "When Facts Are Not Enough." *Science* 360 (2018): 943. https://doi.org/10.1126/science.aau2565.

Hedley, Douglas. "Sophia and the World Soul." In *The Cambridge Companion to Christianity and the Environment*, edited by Alexander J. B. Hampton and Douglas Hedley, 289–302. Cambridge: Cambridge University Press, 2022.

Herman, Judith Lewis. *Trauma and Recovery*. New York: Basic Books, 1992.

Heyse-Moore, Louis. "Does Gaia Experience Trauma?" *European Journal of Ecopsychology* 7 (2022): 75–99.

Hicks, Shari Renée. "A Critical Analysis of Post Traumatic Slave Syndrome: A Multigenerational Legacy of Slavery." Unpublished doctoral thesis (California Institute of Integral Studies, 2015). Accessed 16 August 2024. https://www.proquest.com/dissertations-theses/critical-analysis-post-traumatic-slave-syndrome/docview/1707689965/se-2?accountid=13042.

Hill, Preston. "Does God Need a Body to Keep the Score of Trauma?" *Theological Puzzles* (2021). Accessed 9 January 2021. https://www.theo-puzzles.ac.uk/2021/04/20/phill/.

Hoggard Creegan, Nicola. *Animal Suffering and the Problem of Evil*. New York: Oxford University Press, 2013.

Horrell, David G., and Dominic Coad. "'The Stones Would Cry Out' (Luke 19:40): A Lukan Contribution to a Hermeneutics of Creation's Praise." *Scottish Journal of Theology* 64 (2011): 29–44. https://doi.org/10.1017/S0036930610001043.

Howe, Cymene. "'Timely' Theorizing the Contemporary." *Fieldsights* 2016. Accessed 5 March 2021. https://culanth.org/fieldsights/timely.

Howe, Cymene, and Dominic Boyer. "Death of a Glacier." *Anthropology News* 61 (2020): 16–21.

Hull, John M. "The Broken Body in a Broken World: A Contribution to a Christian Doctrine of the Person from a Disabled Point of View." *Journal of Religion, Disability & Health* 7 (2004): 5–23. https://doi.org/10.1300/J095v07n04_02.

Hunt, Cherryl, David G. Horrell, and Christopher Southgate. "An Environmental Mantra? Ecological Interest in Romans 8:19–23 and a Modest Proposal for Its Narrative Interpretation." *Journal of Theological Studies* 59 (2008): 546–79. https://doi.org/10.1093/jts/fln064.

Hutton, James. *Theory of the Earth; Or an Investigation of the Laws Observable in the Composition, Dissolution, and Restoration of Land Upon the Globe (Transactions of the Royal Society of Edinburgh)*. London: Forgotten Books, [1788] 2007.

International Union of Geological Sciences. "The Anthropocene." (2024). Accessed 6 August 2024. https://www.iugs.org/_files/ugd/f1fc07_40d1a7ed58de458c9f8f24de5e739663.pdf?index=true.

Ions, Rosamund, and Kate Wild. "The Language of Climate Change and Environmental Sustainability." *OED* (2021). Accessed 11 April 2024. https://www.oed.com/discover/the-language-of-climate-change/?tl=true.

IPBES. *Summary for Policymakers of the Global Assessment Report on Biodiversity and Ecosystem Services of the Intergovernmental Science-Policy Platform on Biodiversity and Ecosystem Services*. Edited by S. Díaz, J. Settele, E. S. Brondízio et al. Bonn, Germany: IPBES Secretariat, 2019. https://doi.org/10.5281/zenodo.3553579.

IPCC. "Summary for Policymakers." In *Climate Change 2021: The Physical Science Basis. Contribution of Working Group I to the Sixth Assessment Report of the Intergovernmental Panel on Climate Change*, edited by V. Masson-Delmotte P. Zhai, A. Pirani et al. Cambridge: Cambridge University Press, 2021. https://www.ipcc.ch/report/ar6/wg1/.

Irenaeus. "Against Heresies." In *Ante-Nicene Fathers, Vol. 1*. Edited by Alexander Roberts, James Donaldson, A. Cleveland Coxe, and Kevin Knight. Translated by Alexander Roberts and William Rambaut. Buffalo, NY: Christian Literature Publishing Co., 1885. http://www.newadvent.org/fathers/0103.htm.

Irenaeus. *On the Apostolic Preaching*. Translated by John Behr. Crestwood, NY: St. Vladimir's Seminary Press, 1997.

Irvine, Richard. *An Anthropology of Deep Time: Geological Temporality and Social Life*. New Departures in Anthropology. Cambridge: Cambridge University Press, 2020.

Irvine, Richard. "Seeing Environmental Violence in Deep Time." *Environmental Humanities* 10 (2018): 257–72. https://doi.org/10.1215/22011919-4385562.

Ison, Hilary. "Working with an Embodied and Systemic Approach to Trauma and Tragedy." In *Tragedies and Christian Congregations: The Practical Theology of Trauma*, edited by Megan Warner, Christopher Southgate, Carla A. Grosch-Miller, and Hilary Ison, 47–63. Abingdon: Routledge, 2019.

Jamieson, Dale. *Reason in a Dark Time: Why the Struggle Against Climate Change Failed–and What It Means for Our Future*. Oxford: Oxford University Press, 2014.

Jax, Aurica, and Saskia Wendel. "Introduction." In *Envisioning the Cosmic Body of Christ: Embodiment, Plurality and Incarnation*, edited by Aurica Jax and Saskia Wendel, 1–4. Abingdon: Routledge, 2020.

Jenson, Robert W. "Scripture's Authority in the Church." In *The Art of Reading Scripture*, edited by Ellen F. Davis and Richard B. Hays, 27–37. Grand Rapids, MI: Eerdmans, 2003.

Joerstad, Mari. *The Hebrew Bible and Environmental Ethics: Humans, Non-Humans, and the Living Landscape*. Cambridge: Cambridge University Press, 2019.

Johnson, Elizabeth A. "Animals' Praise of God." *Interpretation: A Journal of Bible and Theology* 73 (2019): 259–71. https://doi.org/10.1177/0020964319838804.

Johnson, Elizabeth A. *Ask the Beasts: Darwin and the God of Love*. London: Bloomsbury, 2014.

Johnson, Elizabeth A. *Creation and the Cross: The Mercy of God for a Planet in Peril*. Maryknoll, NY: Orbis Books, 2018.

Johnson, Elizabeth A. "Jesus and the Cosmos: Soundings in Deep Christology." In *Incarnation: On the Scope and Depth of Christology*, edited by Niels Henrik Gregersen, 133–56. Minneapolis: Fortress Press, 2015.
Jonas, Hans. "Epilogue: The Outcry of Mute Things." In *Mortality and Morality: A Search for the Good After Auschwitz*, edited by Lawrence Vogel, 198–202. Evanston, IL: Northwestern University Press, 1996.
Jones, Serene. *Calvin and the Rhetoric of Piety*, 1st Edition. Columbia Series in Reformed Theology. Louisville, KY: Westminster John Knox Press, 1995.
Jones, Serene. "Hope Deferred: Theological Reflections on Reproductive Loss (Infertility, Stillbirth, Miscarriage)." *Modern Theology* 17 (2001): 227–45. https://doi.org/10.1111/1468-0025.00158.
Jones, Serene. *Trauma and Grace: Theology in a Ruptured World*, 2nd Edition. Louisville, KY: Westminster John Knox Press, 2019.
Jung, Carl Gustav. *Memories, Dreams, Reflections*. New York: Random House, 1961.
Kansteiner, Wulf. "Genealogy of a Category Mistake: A Critical Intellectual History of the Cultural Trauma Metaphor." *Rethinking History* 8 (2004): 193–221. https://doi.org/10.1080/1364252041000168390S.
Kaplan, E. Ann. *Climate Trauma: Foreseeing the Future in Dystopian Film and Fiction*. New Brunswick, NJ: Rutgers University Press, 2016.
Kaplan, E. Ann. "Is Climate-Related Pre-Traumatic Stress Syndrome a Real Condition?" *American Imago* 77 (2020): 81–104. https://doi.org/10.1353/aim.2020.0004.
Kavusa, Kivatsi Jonathan. "Social Disorder and the Trauma of the Earth Community: Reading Hosea 4:1–3 in Light of Today's Crises." *Old Testament Essays* 3 (2016): 481–501. https://doi.org/10.17159/2312-3621/2016/v29n3a8.
Keller, Catherine. *Face of the Deep: A Theology of Becoming*. Abingdon: Routledge, 2003.
Keller, Catherine. *Facing Apocalypse: Climate, Democracy, and Other Last Chances*. Maryknoll, NY: Orbis Books, 2021.
Keller, Catherine. *Intercarnations: Exercises in Theological Possibility*. New York: Fordham University Press, 2017.
Keller, Catherine. *On the Mystery: Discerning Divinity in Process*. Minneapolis: Fortress Press, 2008.
Keller, Catherine. *Political Theology of the Earth: Our Planetary Emergency and the Struggle for a New Public*. Insurrections: Critical Studies in Religion, Politics, and Culture. New York: Columbia University Press, 2018.
Kelly, J. N. D. *Early Christian Doctrines*. London: Adam and Charles Black, 1975.
Keshgegian, Flora A. *Time for Hope: Practices for Living in Today's World*. New York: Continuum, 2006.
Kidd, Erin. "The Violation of God in the Body of the World: A Rahnerian Response to Trauma." *Modern Theology* 35 (2019): 663–82. https://doi.org/10.1111/moth.12484.
Kingsnorth, Paul, and Dougald Hine. "Uncivilisation: The Dark Mountain Manifesto." *The Dark Mountain Project* (2009). Accessed 7 June 2024. https://dark-mountain.net/about/manifesto/.

Kohn, Eduardo. *How Forests Think: Toward an Anthropology Beyond the Human.* Berkeley: University of California Press, 2013.

Komesaroff, Paul, and Ian Kerridge. "A Continent Aflame: Ethical Lessons from the Australian Bushfire Disaster." *Journal of Bioethical Inquiry* 17 (2020): 11–14. https://doi.org/10.1007/s11673-020-09968-9.

Kwok, Pui-lan. "Ecology and Christology." *Feminist Theology* 5 (1989): 113–25. https://doi.org/10.1177/096673509700001508.

Kwok, Pui-lan. "Response to Sallie McFague." In *Christianity and Ecology: Seeking the Well-Being of Earth and Humans,* edited by Dieter T. Hessel and Rosemary Radford Ruether, 47–50. Cambridge, MA: Harvard University Press, 2000.

Lakoff, George, and Mark Johnson. *Metaphors We Live By.* Chicago: University of Chicago Press, 1980.

LaMothe, Ryan. "This Changes Everything: The Sixth Extinction and Its Implications for Pastoral Theology." *Journal of Pastoral Theology* 26 (2016): 178–94. https://doi.org/10.1080/10649867.2016.1275929.

Land, Nick. *Fanged Noumena: Collected Writings 1987–2007,* 5th Edition. Edited by Robin Mackay and Ray Brassier. Falmouth: Urbanomic, 2018.

Lange, Dirk G. *Trauma Recalled: Liturgy, Disruption, and Theology.* Minneapolis: Fortress Press, 2010.

Latour, Bruno. *Facing Gaia: Eight Lectures on the New Climatic Regime.* Translated by Catherine Porter. Cambridge: Polity Press, 2017.

Latour, Bruno. "The Immense Cry Channeled by Pope Francis." *Environmental Humanities* 2 (2016): 251–55. https://doi.org/10.1215/22011919-3664360.

Latour, Bruno. *We Have Never Been Modern.* Translated by Catherine Porter. Cambridge, MA: Harvard University Press, 1993.

Laub, Dori. "Bearing Witness or the Vicissitudes of Listening." In *Testimony: Crises of Witnessing in Literature, Psychoanalysis, and History,* edited by Shoshana Felman and Dori Laub, 57–74. Abingdon: Taylor & Francis, 1992.

Laub, Dori. "Truth and Testimony: The Process and the Struggle." In *Trauma: Explorations in Memory,* edited by Cathy Caruth, 61–75. Baltimore, MD: Johns Hopkins University Press, 1995.

Lear, Jonathan. *Radical Hope: Ethics in the Face of Cultural Devastation.* Cambridge, MA: Harvard University Press, 2006.

Leidenhag, Joanna. *Minding Creation: Theological Panpsychism and the Doctrine of Creation.* T & T Clark Studies in Systematic Theology. London: T & T Clark, 2020.

Leidenhag, Joanna. "Panpsychism and God." *Philosophy Compass* (2022): 1–11. https://doi.org/10.1111/phc3.12889.

Lenton, Timothy, Johan Rockström, Owen Gaffney et al. "Climate Tipping Points—Too Risky to Bet Against." *Nature* 575 (2019): 592–95. https://www.nature.com/articles/d41586-019-03595-0.

Lewis, Simon L., and Mark A. Maslin. "Defining the Anthropocene." *Nature* 519 (2015): 171–80. https://doi.org/10.1038/nature14258.

Liederbach, Mark D. "Stewardship: A Biblical Concept?" In *The Oxford Handbook of the Bible and Ecology*, edited by Hilary Marlow and Mark Harris, 310–23. Oxford: Oxford University Press, 2022.

Linzey, Andrew. *Animal Rites: Liturgies of Animal Care*. London: SCM Press, 1999.

Lloyd, Michael. "The Fallenness of Nature: Three Nonhuman Suspects." In *Finding Ourselves After Darwin: Conversations on the Image of God, Original Sin, and the Problem of Evil*, edited by Stanley P. Rosenberg, 262–79. Grand Rapids, MI: Baker Academic, 2018.

Lopresti-Goodman, Stacy M., Jocelyn Bezner, and Chelsea Ritter. "Psychological Distress in Chimpanzees Rescued from Laboratories." *Journal of Trauma and Dissociation* 16 (2015): 349–66. https://doi.org/10.1080/15299732.2014.1003673.

Lovelock, James. *Gaia: A New Look at Life on Earth*. Oxford: Oxford University Press, 2000.

Lovelock, James. *Gaia: The Practical Science of Planetary Medicine*. London: Gaia, 1991.

Lovelock, James. *The Revenge of Gaia: Why the Earth Is Fighting Back—and How We Can Still Save Humanity*. New York: Penguin, 2007.

Loya, Melissa Tubbs. "Therefore the Earth Mourns: The Grievance of Earth in Hosea 4:1–3." In *Exploring Ecological Hermeneutics*, edited by Norman C. Habel and Peter Trudinger, 53–62. Atlanta: Society of Biblical Literature, 2008.

Luckhurst, Roger. *The Trauma Question*. London: Routledge, 2008.

Luther, Martin. "A Meditation on Christ's Passion." In *The Works of Martin Luther, Volume 42: Devotional Writings, Volume One*. Edited by Martin O. Dietrich. Translated by Martin H. Bertram. Charlottesville, VA: Fortress Press, 1969.

Macy, Joanna. *World as Lover, World as Self*. Berkeley, CA: Parallax Press, 2007.

Malcolm, Hannah. "Grieving the Earth as Prayer: A Wounded Speech That Heals." *Ecumenical Review* 72 (2020): 581–95. https://doi.org/10.1111/erev.12548.

Malm, Andreas, and Alf Hornborg. "The Geology of Mankind? A Critique of the Anthropocene Narrative." *Anthropocene Review* 1 (2014): 62–69. https://doi.org/10.1177/2053019613516291.

Mattes, Eva, and Franco Mattes. *Fukushima Texture Pack*. Creative Capital (2016). Accessed 11 November 2021. https://0100101110101101.org/fukushima-texture-pack/.

Mattes, Eva, and Franco Mattes. *Fukushima Texture Pack*. Creative Capital (2016). Accessed 11 November 2021. https://0100101110101101.org/page/16/.

Mbembe, Achille. *Necropolitics*. Translated by Steve Corcoran. Durham, NC: Duke University Press, 2019.

McCarroll, Pamela R. "Embodying Theology: Trauma Theory, Climate Change, Pastoral and Practical Theology." *Religions* 13 (2022): 1–14.

McCarroll, Pamela R. *Waiting at the Foot of the Cross: Toward a Theology of Hope for Today*. Eugene, OR: Wipf & Stock, 2014.

McDaniel, Jay B. *Of God and Pelicans: A Theology of Reverence for Life*. Louisville, KY: Westminster John Knox Press, 1989.

McFague, Sallie. *The Body of God: An Ecological Theology*. London: SCM Press, 1993.

McFague, Sallie. *The Body of God: An Ecological Theology*. London: SCM Press, 1993.
McFague, Sallie. "An Ecological Christology: Does Christianity Have It?" In *Christianity and Ecology: Seeking the Well-Being of Earth and Humans*, edited by Dieter T. Hessel and Rosemary Radford Ruether, 29–45. Cambridge, MA: Harvard University Press, 2000.
McFague, Sallie. *Models of God: Theology for an Ecological, Nuclear Age*. London: SCM Press, 1987.
McFague, Sallie. *Super, Natural Christians: How We Should Love Nature*. Minneapolis: Fortress Press, 1997.
McPhee, John. *Basin and Range*. New York: Farrar, Straus and Giroux, 1981.
Middleton, Timothy A. "Christic Witnessing: A Practical Response to Ecological Trauma." *Practical Theology* 15 (2022): 420–31. https://doi.org/10.1080/1756073X.2022.2063781.
Middleton, Timothy A. "Christology and the Temporal Trauma of the Anthropocene." In *Rethinking Theology in the Anthropocene*, edited by Andreas Krebs, 63–80. Darmstadt: Wissenschaftliche Buchgesellschaft, 2024.
Midgley, Mary. *Gaia: The Next Big Idea*. London: Demos, 2001. https://doi.org/10.1007/bf00931392.
Miller, Richard W. "Deep Responsibility for the Deep Future." *Theological Studies* 77 (2016): 436–65. https://doi.org/10.1177/0040563916636488.
Miller-McLemore, Bonnie J. "Climate Violence and Earth Justice: A Research Report on Practical Theology's Contributions." *International Journal of Practical Theology* 26 (2022): 329–66. https://doi.org/10.1515/ijpt-2022-0037.
Minns, Denis. *Irenaeus: An Introduction*. London: T & T Clark, 2010.
Moltmann, Jürgen. *The Crucified God: The Cross of Christ as the Foundation and Criticism of Christian Theology*. Translated by R. A. Wilson and John Bowden. London: SCM Press, 1974.
Moltmann, Jürgen. *God in Creation: An Ecological Doctrine of Creation*. Translated by Margaret Kohl. London: SCM Press, 1985.
Moltmann, Jürgen. "Is God Incarnate in All That Is?" In *Incarnation: On the Scope and Depth of Christology*, edited by Niels Henrik Gregersen, 119–32. Minneapolis: Fortress Press, 2015.
Moritz, Joshua M. "Deep Incarnation and the Imago Dei: The Cosmic Scope of the Incarnation in Light of the Messiah as the Renewed Adam." *Theology and Science* 11 (2013): 436–43. https://doi.org/10.1080/14746700.2013.836893.
Morton, Timothy. *Hyperobjects: Philosophy and Ecology After the End of the World*. Minneapolis: University of Minnesota Press, 2013.
Moser, Susanne C. "The Work After 'It's Too Late' (to Prevent Dangerous Climate Change)." *Wiley Interdisciplinary Reviews: Climate Change* 11 (2020): 1–11. https://doi.org/10.1002/wcc.606.
Moss, Candida R. "The Marks of the Nails: Scars, Wounds and the Resurrection of Jesus in John." *Early Christianity* 8 (2017): 48–68. https://doi.org/10.1628/186870317x14876711440088.

Murphy, Nancey. "Science and the Problem of Evil: Suffering as a By-Product of a Finely Turned Cosmos." In *Physics and Cosmology: Scientific Perspectives on the Problem of Natural Evil*, edited by Robert John Russell and William R. Stoeger, 131–52. Vatican City: Vatican Observatory Publications, 2007.

Murray, Michael J. *Nature Red in Tooth and Claw: Theism and the Problem of Animal Suffering*. Oxford: Oxford University Press, 2008.

Murray Parkes, Colin. "Responding to Grief and Trauma in the Aftermath of Disaster." In *Death, Dying, and Bereavement: Contemporary Perspectives, Institutions, and Practices*, edited by Judith Stillion and Thomas Attig, 363–78. New York: Springer, 2014.

Nagel, Thomas. *Mind and Cosmos: Why the Materialist Neo-Darwinian Conception of Nature Is Almost Certainly False*. New York: Oxford University Press, 2012. https://doi.org/10.1093/acprof:oso/9780199919758.001.0001.

Narine, Anil. *Eco-Trauma Cinema*. New York: Routledge, 2015.

Nash, James A. *Loving Nature: Ecological Integrity and Christian Responsibility*. Nashville, TN: Abingdon Press, 1991.

Nazianzen, Gregory. "The First Letter to Cledonius the Presbyter." In *On God and Christ: The Five Theological Orations and Two Letters to Cledonius*. Translated by Lionel Wickham, 155–66. Crestwood, NY: St Vladimir's Seminary Press, 2002.

Nixon, Rob. *Slow Violence and the Environmentalism of the Poor*. Cambridge, MA: Harvard University Press, 2011.

Northcott, Michael S. *The Environment and Christian Ethics*. Cambridge: Cambridge University Press, 1996.

Northcott, Michael S. "Eschatology in the Anthropocene: From the Chronos of Deep Time to the Kairos of the Age of Humans." In *The Anthropocene and the Global Environment Crisis: Rethinking Modernity in a New Epoch*, edited by Clive Hamilton, Christophe Bonneuil, and François Gemenne, 100–111. Abingdon: Routledge, 2015.

Northcott, Michael S. "Religious Traditions and Ecological Knowledge." In *The Cambridge Companion to Christianity and the Environment*, edited by Alexander J. B. Hampton and Douglas Hedley, 231–46. Cambridge: Cambridge University Press, 2022.

O'Donnell, Karen. *Broken Bodies: The Eucharist, Mary, and the Body in Trauma Theology*. London: SCM Press, 2018.

O'Donnell, Karen. "Eucharist and Trauma: Healing in the B/body." In *Tragedies and Christian Congregations: The Practical Theology of Trauma*, edited by Megan Warner, Christopher Southgate, Carla A. Grosch-Miller, and Hilary Ison, 182–93. Abingdon: Routledge, 2019.

O'Donnell, Karen, "The Voices of the Marys: Towards a Method in Feminist Trauma Theologies." In *Feminist Trauma Theologies: Body, Scripture & Church in Critical Perspective*, edited by Karen O'Donnell and Katie Cross, 3–20. London: SCM Press, 2020.

O'Donnell, Karen, and Katie Cross. "Introduction." In *Feminist Trauma Theologies: Body, Scripture & Church in Critical Perspective*, edited by Karen O'Donnell and Katie Cross, xix–xxv. London: SCM Press, 2020.

O'Neill, Saffron J., and Sophie Nicholson-Cole. "'Fear Won't Do It': Promoting Positive Engagement with Climate Change Through Visual and Iconic Representations." *Science Communication* 30 (2009): 355–79. https://doi.org/10.1177/1075547008329201.

Oliver, Kelly. *Witnessing: Beyond Recognition*. Minneapolis: University of Minnesota Press, 2001.

Oord, Thomas Jay. "An Open Theology Doctrine of Creation and Solution to the Problem of Evil." In *Creation Made Free: Open Theology Engaging Science*, edited by Thomas Jay Oord, 28–52. Eugene, OR: Wipf & Stock, 2009.

Orange, Donna. *Climate Crisis, Psychoanalysis and Radical Ethics*. Abingdon: Routledge, 2017.

Orange, Donna. "Traumatized by Transcendence: My Other's Keeper." In *Trauma and Transcendence: Suffering and the Limits of Theory*, 1st Edition, edited by Eric Boynton and Peter Capretto, 70–82. New York: Fordham University Press, 2018. https://doi.org/10.2307/j.ctv19x52c.6.

Ouderkirk, Wayne. "Can Nature Be Evil? Rolston, Disvalue, and Theodicy." *Environmental Ethics* (1999): 135–50. https://doi.org/10.5840/enviroethics199921227.

Page, Ruth. *Ambiguity and the Presence of God*. London: SCM Press, 1985.

Page, Ruth. *God and the Web of Creation*. London: SCM Press, 1996.

Page, Ruth. *The Incarnation of Freedom and Love*. London: SCM Press, 1991.

Peacocke, Arthur. "Biological Evolution: A Positive Theological Appraisal." In *Evolutionary and Molecular Biology: Scientific Perspectives on Divine Action*, edited by Robert John Russell, William R. Stoeger, and Francisco J. Ayala, 357–76. Vatican City: Vatican Observatory Publications, 1998.

Pickstock, Catherine. *Repetition and Identity*. Literary Agenda. Oxford: Oxford University Press, 2013.

Pihkala, Panu. "The Cost of Bearing Witness to the Environmental Crisis: Vicarious Traumatization and Dealing with Secondary Traumatic Stress Among Environmental Researchers." *Social Epistemology* 34 (2020): 86–100. https://doi.org/10.1080/02691728.2019.1681560.

Pihkala, Panu. "Eco-Anxiety, Tragedy, and Hope: Psychological and Spiritual Dimensions of Climate Change." *Zygon* 53 (2018): 545–69. https://doi.org/10.1111/zygo.12407.

Plato. *Plato: Timaeus and Critias*. Edited and translated by Alfred Edward Taylor. Abingdon: Routledge, 2013. https://doi.org/10.4324/9780203101360.

Pope Francis. *Encyclical Letter Laudato Si' of the Holy Father Francis: On Care for Our Common Home*. Vatican City: Vatican Press, 2015. https://www.vatican.va/content/dam/francesco/pdf/encyclicals/documents/papa-francesco_20150524_enciclica-laudato-si_en.pdf.

Pound, Marcus. *Theology, Psychoanalysis and Trauma*. London: SCM Press, 2007.
Poussaint, Alvin F., and Amy Alexander. *Lay My Burden Down: Suicide and the Mental Health Crisis Among African Americans*. Boston: Beacon Press, 2000.
Radford Ruether, Rosemary. *Sexism and God-Talk*. London: SCM Press, 1983.
Rae, Eleanor. "Response to Mark I. Wallace: Another View of the Spirit's Work." In *Christianity and Ecology: Seeking the Well-Being of Earth and Humans*, edited by Dieter T. Hessel and Rosemary Radford Ruether, 73–82. Cambridge, MA: Harvard University Press, 2000.
Rahner, Karl. *Mission and Grace: Essays in Pastoral Theology II*. Translated by Cecily Hastings and Richard Strachan. London: Sheed & Ward, 1963.
Rambo, Shelly. "Changing the Conversation: Theologizing War in the Twenty-First Century." *Theology Today* 69 (2013): 441–62. https://doi.org/10.1177/00405736 12463035.
Rambo, Shelly. "Introduction." In *Post-Traumatic Public Theology*, edited by Stephanie N. Arel and Shelly Rambo, 1–21. Cham, Switzerland: Palgrave Macmillan, 2016.
Rambo, Shelly. *Resurrecting Wounds: Living in the Afterlife of Trauma*. Waco, TX: Baylor University Press, 2017.
Rambo, Shelly. "Saturday in New Orleans: Rethinking the Holy Spirit in the Aftermath of Trauma." *Review & Expositor* 105 (2008): 229–44. https://doi.org/10 .1177/003463730810500206.
Rambo, Shelly. *Spirit and Trauma: A Theology of Remaining*. Louisville, KY: Westminster John Knox Press, 2010.
Rappmann, Susanne. "The Disabled Body of Christ as a Critical Metaphor—Towards a Theory." *Journal of Religion, Disability & Health* 7 (2004): 25–40. https://doi.org/10.1300/J095v07n04_03.
Reid, Duncan. "Enfleshing the Human: An Earth-Revealing, Earth-Healing Christology." In *Earth Revealing—Earth Healing: Ecology and Christian Theology*, edited by Denis Edwards, 69–83. Collegeville, MN: The Liturgical Press, 2001.
Rensberger, David. "Ecological Use of the Psalms." In *The Oxford Handbook of The Psalms*. edited by William P. Brown, 608–19. Oxford Handbooks. Oxford: Oxford University Press, 2014.
Richardson, Michael. "Climate Trauma, or the Affects of the Catastrophe to Come." *Environmental Humanities* 10 (2018): 1–19. https://doi.org/10.1215/22011919-4385444.
Richardson, Michael. *Nonhuman Witnessing: War, Data, and Ecology After the End of the World*. Durham, NC: Duke University Press, 2024.
Ricœur, Paul. *The Rule of Metaphor: The Creation of Meaning in Language*. Translated by Robert Czerny, Kathleen McLaughlin, and John Costello. London: Routledge, 2003.
Rivera, Mayra. *Poetics of the Flesh*. Durham, NC: Duke University Press, 2015.
Rolston, Holmes, III. "Disvalues in Nature." *The Monist* 75 (1992): 250–78. https://doi .org/10.5840/monist199275218.

Rolston, Holmes, III. *Genes, Genesis and God: Values and Their Origins in Natural and Human History*. Cambridge: Cambridge University Press, 1999.

Rolston, Holmes, III. "Redeeming a Cruciform Nature." *Zygon* 53 (2018): 739–51. http://doi.wiley.com/10.1111/zygo.12428.

Rolston, Holmes, III. *Science & Religion: A Critical Survey*. Philadelphia: Templeton Foundation Press, 2006.

Roszak, Theodore. *Person/Planet: The Creative Disintegration of Industrial Society*. London: Granada, 1981.

Roszak, Theodore. *The Voice of the Earth: An Exploration of Ecopsychology*. New York: Simon & Schuster, 1992.

Rothberg, Michael. *Multidirectional Memory: Remembering the Holocaust in the Age of Decolonization*. Cultural Memory in the Present. Stanford, CA: Stanford University Press, 2009.

Rothberg, Michael. "Preface: Beyond Tancred and Clorinda—Trauma Studies for Implicated Subjects." In *The Future of Trauma Theory: Contemporary Literary and Cultural Criticism,* edited by Gert Buelens, Sam Durrant, and Robert Eaglestone, xi–xviii. Abingdon: Routledge, 2013.

Rothschild, Babette. *The Body Remembers: The Psychophysiology of Trauma and Trauma Treatment*. New York: W. W. Norton, 2000.

Rots, Aike P., and Nhung Lu Rots. "When Gods Drown in Plastic: Vietnamese Whale Worship, Environmental Crises, and the Problem of Animism." *Environmental Humanities* 15 (2023): 8–29. https://doi.org/10.1215/22011919-10745957.

Rubenstein, Mary-Jane. "Afterword." In *Trauma and Transcendence: Suffering and the Limits of Theory,* 1st Edition, edited by Eric Boynton and Peter Capretto, 283–94. New York: Fordham University Press, 2018. https://doi.org/10.2307/j.ctv19x52c.16.

Rubenstein, Mary-Jane. *Pantheologies: Gods, Worlds, Monsters*. New York: Columbia University Press, 2018.

Sabini, Meredith, ed. *The Earth Has A Soul: C. G. Jung on Nature, Technology & Modern Life*. Berkeley, CA: North Atlantic Books, 2016.

Santmire, H. Paul. "Behold the Lilies: Martin Buber and the Contemplation of Nature." *Dialog* 57 (2018): 18–22. https://doi.org/10.1111/dial.12372.

Santmire, H. Paul. "I-Thou, I-It, and I-Ens." *The Journal of Religion* 48 (1968): 260–73. https://doi.org/10.1086/486128.

Scarsella, Hilary Jerome. "Trauma and Theology." In *Trauma and Transcendence: Suffering and the Limits of Theory,* 1st Edition, edited by Eric Boynton and Peter Capretto, 256–82. New York: Fordham University Press, 2018. https://doi.org/10.2307/j.ctv19x52c.15.

Schneider, Laurel C. "Promiscuous Incarnation." In *The Embrace of Eros: Bodies, Desires, and Sexuality in Christianity,* edited by Margaret D. Kamitsuka, 231–46. Minneapolis: Fortress Press, 2010.

Schreiner, Susan E. *The Theater of His Glory: Nature and the Natural Order in the Thought of John Calvin*. Studies in Historical Theology. Durham, NC: Labyrinth Press, 1991.
Schüssler Fiorenza, Elisabeth. "Proclaimed by Women: The Execution of Jesus and the Theology of the Cross." In *Jesus: Miriam's Child, Sophia's Prophet: Critical Issues in Feminist Christology*, 105–39. London: Bloomsbury T & T Clark, 2015.
Sender, Ron, Shai Fuchs, and Ron Milo. "Are We Really Vastly Outnumbered? Revisiting the Ratio of Bacterial to Host Cells in Humans." *Cell* 164 (2016): 337–40. https://doi.org/10.1016/j.cell.2016.01.013.
Sharpe, Christina Elizabeth. *In the Wake: On Blackness and Being*. Durham, NC: Duke University Press, 2016.
Sideris, Lisa H. "Grave Reminders: Grief and Vulnerability in the Anthropocene." *Religions* 11 (2020): 1–16. https://doi.org/10.3390/rel11060293.
Skrimshire, Stefan. "Anthropocene Fever: Memory and the Planetary Archive." In *Religion and the Anthropocene*, edited by Celia Deane-Drummond, Sigurd Bergmann, and Markus Vogt, 138–54. Eugene, OR: Cascade Books, 2017.
Skrimshire, Stefan. "Deep Time and Secular Time: A Critique of the Environmental 'Long View.'" *Theory, Culture and Society* 36 (2019): 63–81. https://doi.org/10.1177/0263276418777307.
Skrimshire, Stefan. "Eternal Return of the Apocalypse." In *Future Ethics: Climate Change and Apocalyptic Imagination*, edited by Stefan Skrimshire, 219–41. London: Continuum, 2010.
Smesler, Neil J. "Psychological Trauma and Cultural Trauma." In *Cultural Trauma and Collective Identity*, edited by Jeffrey C. Alexander, Ron Eyerman, Bernhard Giesen, Neil J. Smesler, and Piotr Sztompka, 31–59. Berkeley: University of California Press, 2004.
Smith, John E. "Time, Times, and the 'Right Time'; Chronos and Kairos." *The Monist* 53 (1969): 1–13. https://doi.org/10.5840/monist196953115.
Sobrino, Jon. *Christ the Liberator: A View from the Victims*. Translated by Paul Burns. Maryknoll, NY: Orbis Books, 2001.
Sobrino, Jon. *Jesus the Liberator: A Historical-Theological Reading of Jesus of Nazareth*. Translated by Paul Burns and Francis McDonagh. Maryknoll, NY: Orbis Books, 1993.
Sollereder, Bethany N. "Compassionate Theodicy: A Suggested Truce Between Intellectual and Practical Theodicy." *Modern Theology* 37 (2021): 382–95. https://doi.org/10.1111/moth.12688.
Sollereder, Bethany N. *God, Evolution, and Animal Suffering: Theodicy Without a Fall*. Abingdon: Routledge, 2019.
Solnit, Rebecca. *Hope in the Dark: Untold Histories, Wild Possibilities*. Edinburgh: Canongate Books, 2016.
Sorrell, Roger D. *St. Francis of Assisi and Nature: Tradition and Innovation in Western Christian Attitudes Toward the Environment*. Oxford: Oxford University Press, 2009).

Southgate, Christopher. "Depth, Sign and Destiny: Thoughts on Incarnation." In *Incarnation: On the Scope and Depth of Christology*, edited by Niels Henrik Gregersen, 203–24. Minneapolis: Fortress Press, 2015.

Southgate, Christopher. "Divine Glory in a Darwinian World." *Zygon* 49 (2014): 784–807. https://doi.org/10.1111/zygo.12126.

Southgate, Christopher. "Does God's Care Make Any Difference? Theological Reflection on the Suffering of God's Creatures." In *Christian Faith and the Earth: Current Paths and Emerging Horizons in Ecotheology*, edited by Ernst M. Conradie, Sigurd Bergmann, Celia Deane-Drummond, and Denis Edwards, 97–114. New York: Bloomsbury T & T Clark, 2014.

Southgate, Christopher. *The Groaning of Creation: God, Evolution, and the Problem of Evil*. Louisville, KY: Westminster John Knox Press, 2008.

Southgate, Christopher. *Theology in a Suffering World: Glory and Longing*. Cambridge: Cambridge University Press, 2018. https://doi.org/10.1017/9781316599945.

St. Columban's Mission Society. "Stations of the Forests." St. Columban's Mission Society (2011), 1–28. Accessed 9 March 2021. https://www.columban.org.au/assets/files/cmi/columban stations of the forest booklet.pdf.

Steenberg, M. C. *Irenaeus on Creation: The Cosmic Christ and the Saga of Redemption*. Leiden: Brill, 2008.

Steffen, Will, Johan Rockström, Katherine Richardson, Timothy M. Lenton, Carl Folke, and Diana Liverman. "Trajectories of the Earth System in the Anthropocene." *PNAS* 115 (2018): 8252–59. https://doi.org/10.1073/pnas.1810141115.

Stein, Leslie. "Global Warming: Inaction, Denial and Psyche." In *Environmental Disasters and Collective Trauma*, edited by Nancy Cater and Stephen Foster, 23–46. New Orleans: Spring Journal, 2012.

Stephens, Kevin U., David Grew, Karen Chin et al. "Excess Mortality in the Aftermath of Hurricane Katrina: A Preliminary Report." *Disaster Medicine and Public Health Preparedness* 1 (2007): 15–20.

Stone, Selina. "Spirit for the Oppressed? Pentecostalism, the Spirit and Black Trauma." In *Bearing Witness: Intersectional Perspectives on Trauma Theology*, edited by Karen O'Donnell and Katie Cross, 58–76. London: SCM Press, 2022.

Strawn, Brent A. "Trauma, Psalmic Discourse, and Authentic Happiness." In *Bible Through the Lens of Trauma*, edited by Elizabeth Boase and Christopher G. Frechette, 143–60. Atlanta: SBL Press, 2016.

Surin, Kenneth. *Theology and the Problem of Evil*. Oxford: Blackwell, 1986.

Swain, Storm. "Climate Change and Pastoral Theology." In *T & T Clark Handbook of Christian Theology and Climate Change*, edited by Ernst M. Conradie and Hilda P. Koster, 615–26. London: T & T Clark, 2020.

Swinton, John. *Raging with Compassion: Pastoral Responses to the Problem of Evil*. Grand Rapids, MI: Eerdmans, 2007.

Szerszynski, Bronislaw. "How the Earth Remembers and Forgets." In *Political Geology: Active Stratigraphies and the Making of Life*, edited by Adam Bobbette and Amy Donovan, 219–36. Cham, Switzerland: Palgrave Macmillan, 2018. https://doi.org/10.1007/978-3-319-98189-5_8.

Tanner, Kathryn. *Christ the Key*. Cambridge: Cambridge University Press, 2010.

Taylor, Bron Raymond. *Dark Green Religion: Nature Spirituality and the Planetary Future*. Berkeley: University of California Press, 2010.

Taylor, Mark C., and Mark Tansey. *The Picture in Question: Mark Tansey and the Ends of Representation*. Chicago: University of Chicago Press, 1999.

Teilhard de Chardin, Pierre. *Hymn of the Universe*. New York: Harper & Row, 1965.

Theokritoff, Elizabeth. "Creation and Salvation in Orthodox Worship." *Ecotheology* 10 (2001): 97–108. https://doi.org/10.1558/ecotheology.v6i1.97.

Thompson, Deanna A. *Glimpsing Resurrection: Cancer, Trauma, and Ministry*. Louisville, KY: Westminster John Knox Press, 2018.

Tilley, Terrence W. *The Evils of Theodicy*. Washington, D.C.: Georgetown University Press, 1991.

Tilling, Chris. "Paul, Christ, and Narrative Time." In *Christ and the Created Order*, edited by Andrew B. Torrance and Thomas H. McCall, 151–66. Grand Rapids, MI: Zondervan, 2018.

Tinker, George "Tink." "The Stones Shall Cry Out: Consciousness, Rocks, and Indians." *Wicazo Sa Review* 19 (2004): 105–25. https://www.jstor.org/stable/1409501.

Tonstad, Sigve K. *The Letter to the Romans: Paul Among the Ecologists*. Sheffield: Sheffield Phoenix Press, 2016.

Tumminio Hansen, Danielle. "The Body of God, Sexually Violated: A Trauma-Informed Reading of the Climate Crisis." *Religions* 13 (2022): 1–12. https://doi.org/10.3390/rel13030249.

Tumminio Hansen, Danielle. "Remembering Rape in Heaven: A Constructive Proposal for Memory and the Eschatological Self." *Modern Theology* 37 (2021): 662–78. https://doi.org/10.1111/moth.12651.

Turia, Tariana. "Tariana Turia's Speech Notes." *Speech to NZ Psychological Society Conference* (2000). Accessed 14 August 2024. http://www.converge.org.nz/pma/tspeech.htm.

Tylor, Edward Burnett. *Religion in Primitive Culture*. New York: Harper & Row, [1871] 1958.

Uehlinger, Christoph. "The Cry of the Earth? Biblical Perspectives on Ecology and Violence." In *Ecology and Poverty: Cry of the Earth, Cry of the Poor*, edited by Leonardo Boff and Virgil Elizondo, 41–57. London: SCM Press, 1995.

Urbaniak, Jakub. "Extending and Locating Jesus's Body: Toward a Christology of Radical Embodiment." *Theological Studies* 80 (2019): 774–97. https://doi.org/10.1177/0040563919874520.

van der Kolk, Bessel A. *The Body Keeps the Score: Brain, Mind, and Body in the Healing of Trauma*. London: Penguin Books, 2015.

van Deusen Hunsinger, Deborah. *Bearing the Unbearable: Trauma, Gospel, and Pastoral Care*. Grand Rapids, MI: Eerdmans, 2015.

Verlie, Blanche. *Learning to Live with Climate Change: From Anxiety to Transformation*, Routledge Focus on Environment and Sustainability. London: Routledge, 2022.

Veron, J. E. N. "Mass Extinctions and Ocean Acidification: Biological Constraints on Geological Dilemmas." *Coral Reefs* 27 (2008): 459–72. https://doi.org/10.1007/s00338-008-0381-8.

Vogt, Markus. *Christian Environmental Ethics: Foundations and Central Challenges*. Translated by Gary Slater. Paderborn: Brill Schöningh, 2023.

Voosen, Paul. "Ice Shelf Holding Back Keystone Antarctic Glacier Within Years of Failure." *Science* 374 (2021): 1420–21. https://doi.org/10.1126/science.acz9833.

Wainwright, Elaine Mary. *Habitat, Human, and Holy: An Eco-Rhetorical Reading of the Gospel of Matthew*. Sheffield: Sheffield Phoenix Press, 2016.

Wallace, Mark I. "Christian Animism, Green Spirit Theology, and the Global Crisis Today." *Journal of Reformed Theology* 6 (2012): 216–33. https://doi.org/10.1057/978 1137268990_15.

Wallace, Mark I. "Elegy for a Lost World." In *Post-Traumatic Public Theology*, edited by Stephanie N. Arel and Shelly Rambo, 135–54. Cham, Switzerland: Palgrave Macmillan, 2016.

Wallace, Mark I. "Even Rocks Are Alive: Christian Animist Disruptions of the Species Divide." In *Taking a Deep Breath for the Story to Begin. . .*, edited by Ernst M. Conradie and Pan-Chiu Lai, 241–58. Cape Town: AOSIS, 2021.

Wallace, Mark I. *Fragments of the Spirit: Nature, Violence, and the Renewal of Creation*. New York: Continuum, 1996.

Wallace, Mark I. "The Lord God Bird: Avian Divinity, Neo-Animism, and the Renewal of Christianity at the End of the World." In *Encountering Earth: Thinking Theologically With a More-Than-Human World*, edited by Trevor Bechtel, Matthew Eaton, and Timothy Harvie, 210–26. Eugene, OR: Cascade Books, 2018.

Wallace, Mark I. *When God Was A Bird*. New York: Fordham University Press, 2019.

Wallace, Mark I. "The Wounded Spirit as the Basis for Hope in an Age of Radical Ecology." In *Christianity and Ecology: Seeking the Well-Being of Earth and Humans*, edited by Dieter T. Hessel and Rosemary Radford Ruether, 51–72. Cambridge, MA: Harvard University Press, 2000.

Walsh, Brian J., Marianne B. Karsh, and Nik Ansel. "Trees, Forestry, and the Responsiveness of Creation." *Cross Currents* 44 (1994): 149–62. https://www.jstor.org/stable/24460092.

Wang, Susie, Adam Corner, Daniel Chapman, and Ezra Markowitz. "Public Engagement with Climate Imagery in a Changing Digital Landscape." *Wiley Interdisciplinary Reviews: Climate Change* 9 (2018): 1–18. https://doi.org/10.1002/wcc.509.

Ward, Graham. *Christ and Culture*. Malden, MA: Blackwell Publishers, 2005.

Ward, Graham. "The Displaced Body of Jesus Christ." In *Radical Orthodoxy: A New Theology*, edited by John Milbank, Catherine Pickstock, and Graham Ward, 163–81. London: Routledge, 1998.

Ward, Graham. *The Politics of Discipleship: Becoming Postmaterial Citizens*. London: SCM Press, 2009.
Warner, Megan. "Teach to Your Daughters a Dirge: Revisiting the Practice of Lament in the Light of Trauma Theory." In *Tragedies and Christian Congregations: The Practical Theology of Trauma*, edited by Megan Warner, Christopher Southgate, Carla A. Grosch-Miller, and Hilary Ison, 167–81. Abingdon: Routledge, 2019.
Warner, Megan. "Trauma Through the Lens of the Bible." In *Tragedies and Christian Congregations: The Practical Theology of Trauma*, edited by Megan Warner, Christopher Southgate, Carla A. Grosch-Miller, and Hilary Ison, 81–91. Abingdon: Routledge, 2019.
Warner, Megan, Christopher Southgate, Carla A. Grosch-Miller, and Hilary Ison, eds. *Tragedies and Christian Congregations: The Practical Theology of Trauma*. Abingdon: Routledge, 2019.
Westhelle, Vítor. *Eschatology and Space: The Lost Dimension in Theology Past and Present*. New York: Palgrave Macmillan, 2012.
Whitaker, Maja. "The Wounds of the Risen Christ: Evidence for the Retention of Disabling Conditions in the Resurrection Body." *Journal of Disability and Religion* (2021): 1–14. https://doi.org/10.1080/23312521.2021.2016547.
White, Benjamin. "States of Emergency: Trauma and Climate Change." *Ecopsychology* 7 (2015): 192–97. https://doi.org/10.1089/eco.2015.0024.
White, Lynn. "The Historical Roots of Our Ecologic Crisis." *Science* 155 (1967): 1203–7. https://doi.org/10.1126/science.155.3767.1203.
White, Robert S. *Who Is to Blame? Disasters, Nature and Acts of God*. Oxford: Monarch Books, 2014.
Whyte, Kyle. "Against Crisis Epistemology." In *Routledge Handbook of Critical Indigenous Studies*, edited by Brendan Hokowhitu, Aileen Moreton-Robinson, Linda Tuhiwai-Smith, Chris Andersen, and Steve Larkin, 52–64. London: Routledge, 2020. https://doi.org/10.4324/9780429440229-6.
Widdicombe, Peter. "The Wounds and the Ascended Body: The Marks of Crucifixion in the Glorified Christ from Justin Martyr to John Calvin." *Laval Théologique et Philosophique* 59 (2003): 137–54. https://doi.org/10.7202/000793ar.
Wiesel, Elie. *Night*. Translated by Marion Wiesel. New York: Hill and Wang, 1958.
Wiles, Maurice F. "The Unassumed Is the Unhealed." *Religious Studies* 4 (1968): 47–56. https://doi.org/10.1017/S0034412500003383.
Williams, Delores S. *Sisters in the Wilderness: The Challenge of Womanist God-Talk*. Maryknoll, NY: Orbis Books, 1993.
Woodbury, Zhiwa. "Climate Trauma: Toward a New Taxonomy of Trauma." *Ecopsychology* 11 (2019): 1–8. https://doi.org/10.1089/eco.2018.0021.
Woolbright, Lauren. "Wounded Planet, Wounded People: The Possibility of Ecological Trauma." Unpublished master's thesis (Clemson University, 2011). Accessed 28 October 2018. https://www.proquest.com/docview/881256145.

Wurst, Shirley. "Retrieving Earth's Voice in Jeremiah: An Annotated Voicing of Jeremiah 4." In *The Earth Story in the Psalms and the Prophets*, edited by Norman C. Habel, 172–84. Sheffield: Sheffield Academic Press, 2001.

Wyman, Jason A. "Interpreting the History of the Workgroup on Constructive Theology." *Theology Today* 73 (2017): 312–24. https://doi.org/10.1177/0040573616669565.

Yusoff, Kathryn. *A Billion Black Anthropocenes or None*. Minneapolis: University of Minnesota Press, 2018.

Zylinska, Joanna. *Minimal Ethics for the Anthropocene*. Ann Arbor, MI: Open Humanities Press, 2014.

Index

Abbott, Roger, 165n36
Abel. *See* Cain and Abel
accompaniment: divine, 11, 87–88, 101, 119, 130; human 9–10, 48, 50, 57, 62, 122
activism, 18, 42, 61, 138
Adam, 45, 105, 119–20, 152n66, 197n73, 198n78
adamah, 198n85
after-living, 132
Airenti, Gabriella, 38–39
Alaimo, Stacy, 138, 204n50
Alexander, Jeffrey, 171n8
Ambrose, 105
American Psychiatric Association, 61
Amos, 154n5
androcentrism, 102
anima mundi. *See* world soul
animal ethology, 162–63n124
animal rights, 61
animism, 25–26, 29, 30–36, 38–39, 65, 160n97; biblical, 19, 22–23, 31, 33; Christian, 15, 31, 34–35, 159n71, 159n76; Indigenous, 33; new, 32–34
annunciation-incarnation event, 139
Anselm, 97
Antarctic ice sheets, 96, 113
Anthropocene, 6, 17, 110–27, 192n6, 201n109; ethics in the, 48, 137, 166n43, 204n45; geology of the, 6, 113, 124, 149n37, 191n6, 193n21. *See also* scars (Anthropocene)
anthropocentrism, 12, 37–38, 45, 60, 84, 102, 120, 152n65, 170n3
anthropomorphism, 16, 19–25, 29, 36–40, 60, 65, 78, 95, 129, 159n74, 161n108, 162n115, 162n117, 162n124, 173n47, 184n10; biblical, 20–23, 31, 71; Pope Francis on, 23–25; imagination and, 20, 38, 40, 129; projection and, 37–38, 40, 161n109, 180n65. *See also* language (anthropomorphic), metaphor (anthropomorphic), nature (anthropomorphisms of), psychology (and anthropomorphism)
apocalypse, 41, 72, 74, 133, 196n63, 204n45
Arendt, Hannah, 166n43
Athanasius, 26
atonement, 62, 106, 153n68, 182n99
Augustine, 186n25, 194n37
Auschwitz, 103, 148n27
Australian wildfires, 41–42, 45–47, 57

Baldwin, Jennifer, 69
Balthasar, Hans Urs von, 79, 115, 151n52, 177n33, 178n37
Bauckham, Richard, 106–7, 161n108, 162n114, 190n86
bearing witness, 6, 11, 17–18, 74–92, 93–94, 99–109, 111, 118–27, 131, 133, 136–38, 140–41, 175n71, 176n9, 177n33, 179n47, 180n73, 204n49
Bearing Witness, 62, 64
Bendell, Jem. *See* Deep Adaptation
Bennett, Jane, 33, 38
Beste, Jennifer, 150n43
Betcher, Sharon, 96, 194n39
Bible. *See* Hebrew Bible, New Testament
biodiversity, 1, 3–4, 42, 44, 46, 64, 113

Black Summer bushfires. *See* Australian wildfires
body: of Christ, x–xi, 11, 80, 84, 90, 97–98, 100, 104–8, 111, 116, 122–26, 132, 137–38, 151n54, 152n60, 169n87, 175n71, 178n41, 183n110, 199n95; of God, 53–54, 58, 79, 91, 169n83, 183n111; in heaven, 123–24; the human, 25, 53, 60, 94. See also *The Body Keeps the Score*, *The Body of God*, *Broken Bodies*, Earth (body of the), flesh (versus bodies)
The Body Keeps the Score, 5–6, 101
The Body of God, 53–54, 169n83
Boff, Leonardo, 156n29
Bonaventure, 90, 183n106
Boynton, Eric, 148n28, 166n46
Bradshaw, Gay, 61
Broken Bodies, 69
Buber, Martin, 35–36, 38, 161n104, 162n117
Buffalo Creek, 67
bushfires. *See* Australian wildfires

Cadwallader, Alan, 72
Cain and Abel, 159n75
Calvin, John, 123, 125, 170n3. *See also* Rambo (on Calvin)
Cannon, Katie, 93, 95, 108
Canticle of the Creatures, 23–24
Capretto, Peter, 166n46
Caravaggio, x, 199n96
carbon dioxide, 28, 65, 113
Caruth, Cathy, 5, 49, 111
Chalcedon, council of, 55, 159n77, 185n22
Chopp, Rebecca, 177n17
Christ, 10–18, 54–55, 86–92, 94, 97–109, 115–26, 129–30, 137–38, 140; ascension body of, 111, 122–26, 132; flesh of, 12, 79–80, 94, 96–102, 108–9, 122, 125–26, 130, 136–37, 191n98, 199n96; genealogies of, 11, 121–22, 198n85; as subject of ecological trauma, 26; wounds of, x–xi, 63, 83, 100, 122–24, 126, 129, 199n95. *See also* body (of Christ), Christic witnessing, creation (Christ and), Jesus, psychology (of Christ), representation (by Christ), scars (on Christ's body), solidarity (of Christ)
Christian animism. *See* animism (Christian)
Christic witnessing, 13, 16–17, 74–92, 98–103, 106–9, 111, 114–27, 130, 132–33, 136–38, 141
Christology, 10–13, 15–17, 55–56, 79–84, 89–91, 96, 100, 105, 111, 114–22, 126, 130, 151n58, 159n77, 177n30, 185n20, 197n71, 200n107. *See also* cosmic Christology, mirroring (Christology of)

chronos, 196n63. *See also* incarnation (and chronocentrism)
Chryssavgis, John, 205n61
church, 116–117, 138–40, 178n41, 183n110
climate change: adaptation to, 17, 131–33; anthropogenic, 1, 46–47, 66, 75, 112–113; and the climate system, 6, 28, 112; and colonialism, 14; as crisis, 2, 42–44, 51; ethics of, 46–49, 166n44; impacts of, 41, 66, 192n14; living with, 78, 133; mental health and, 2, 146n6; mitigation of, 17, 131–33; as traumatic, 3, 67, 112. *See also* global warming, trauma (climate)
Cole-Turner, Ronald, 186n25
collective trauma. *See* trauma (collective)
Collins, Natalie, 149n31
colonialism, 13–15, 29, 33, 38–39, 165n36, 205n60
Colossians, 195n56
communication: breakdown of, 3, 5, 16, 30, 75, 98–99; haptic, 100–1; impossibility of, 5, 101, 148n27; rupture of, 2–7, 16–17, 64, 70, 74–92, 94, 130, 138, 148n24. *See also* Earth (communication of the)
companionship, of creation, 17, 87–88, 91, 181n86, 181n90
Cone, James, 14–15, 103–4, 175n71
Conradie, Ernst, 107–8, 123–24, 190n92
consciousness: distributed, 28, 30, 32, 39; interconnected, 30–31, 39; of matter, 27–28; of rocks, 27, 34, 155n14; of trees, 35. *See also* Earth (consciousness of the)
consolation, 10, 48, 57, 130
Copeland, Rebecca, 70–71
coping, 1, 9, 17, 49, 131–33, 141
co-redeemers, 167n47, 182n98
cosmic Christology, 11–13, 15, 81, 102–3, 106, 185n20, 205n60
cosmopsychism, 27–28, 31, 39
co-suffering, 17, 88–91, 101, 182n93, 182n95. *See also* God (as co-sufferer), salvation (co-suffering and)
Craps, Stef, 13–14, 62
creation: Christ and, 13, 87, 101–8, 121, 197n75; doctrine of, 9, 183n102; groaning of, 21–22, 155n18, 156n27; panpsychism and, 27; praise of, 21, 155n19, 161n108, 201n109; recapitulation of, 119–20; season of, 139. *See also* cruciform (creation)
cross: mirrored, 83–84; stations of the, 139. See also *The Cross and the Lynching Tree*, violence (of the cross)
Cross, Katie, 7–8, 62, 64

INDEX

The Cross and the Lynching Tree, 15, 103–4
crucifixion earthquake, 72, 84, 145n1
cruciform: creation, 17, 85–87, 119–20, 133, 181n75, 197n75; witnessing, 83–85, 89–91, 139, 180n69, 180n73
cry: of dereliction, 71–72, 84–85; of the Earth, 16, 23–25, 36, 39, 60, 75, 84, 156n27, 156n29, 158n66, 159n75; of the land, 156n24; of the nonhuman, 20–21, 201n109; of the poor, 23–24, 39, 156n29, 158n66
cultural trauma. *See* trauma (cultural)
culture. *See* nature/culture binary

Daly, Mary, 152n67
dark green religion, 156n35
Dark Mountain Project, 132–33
Deane-Drummond, Celia, 47, 49, 86, 102–3, 106, 133, 162n124, 178n37, 185n20
decolonization, 2, 13–15, 33
Deep Adaptation, 132–33, 201nn4–6
deep incarnation, 11–12, 17, 93–109, 116–21, 126, 130, 151n59, 152n65, 185n20, 187n32, 191n97
deep time. *See* time (deep)
Deepwater Horizon oil spill, 116
deforestation, 6, 29, 46, 48, 79, 96, 106, 112, 125, 165n36, 180n69
deicide, 57, 91, 170n95
De La Torre, Miguel, 134
disability. *See* theology (disability)
disasters, natural, 2, 4, 20, 45, 66–67, 106, 112, 120, 139, 165n36, 172n36, 173n45; technological, 173n45
disvalue, in nature, 44, 164n22
divinity. *See* accompaniment (divine), companionship, co-suffering, immanence (divine), impassibility, memory (divine), solidarity (divine)
doctrine. *See* Christology, creation, ecclesiology, impassibility, pneumatology, salvation, sin, Trinity
Doehring, Carrie, 150n43
Doubting Thomas, 100–1, 122–23, 137; Tansey's painting of, ix–xi, 129. *See also* Rambo (on Doubting Thomas)

Earth: abuse of the, 20–21, 35, 53, 55, 169n87; activity of the, 27, 29, 37, 72, 156n30, 161n110; body of the, 6, 110, 125; communication of the, 22, 37–40, 157n51; consciousness of the, 3, 15, 19, 22, 25–33, 39; emotion of the, 21, 29, 159n75; history of the, 110, 114; life on, 28, 49, 54, 121; memory of the, 193n25; mourning of the, 20, 71–72, 155n18, 174n66; representation of the, 138; subjectivity of the, 15, 21, 31–32, 36, 39, 156n30; suffering of the, 9, 24, 26, 35, 72, 84–85, 129, 157n51, 169n88, 180n73; system, 3, 8, 26, 65, 95, 111, 113; voice of the, 22, 37, 39, 40, 72, 155n18. *See also* cry (of the Earth), Earth Bible Project, Earth system science, Earth systems stress trauma, flesh (of the Earth), Mother Earth, trauma (Earth), *The Voice of the Earth*
Earth Bible Project, 22, 36, 72
Earth system science, 28, 31
Earth systems stress trauma, 28
earthquakes, ix–xi, 46, 69, 93, 145n3, 165n36. *See also* crucifixion earthquake, Haitian earthquake
Easter Sunday, 79, 115
Eastern Orthodoxy. *See* theology (Eastern Orthodox)
Eaton, Matthew, 16, 51, 56–58, 91, 105, 124, 152n65, 161n110, 169n91, 170n95
Ecclesiastes, 196n63
ecclesiology, 62, 140
ecoanxiety, 2
ecocide, 57, 91, 170n95
ecofeminism, 23
ecojustice principles, 22. *See also* Earth Bible Project
ecological: breakdown, 2, 42, 48, 57, 59, 77; conversion, 71; despair, 2; destruction, 2, 14, 42, 44, 51, 111, 130, 141; grief, 2; hermeneutics, 22, 71; pneumatology, 54–55; sin, 45, 164n25. *See also* flesh (ecological), theology (ecological)
ecopsychology, 15, 25, 29–30, 35, 39, 158n63
ecosystems, 1–2, 4, 14, 40, 49, 70, 113, 158n52
ecotrauma, Keller on, 41, 52–53
Edwards, Denis, 17, 82, 87–89, 97, 119–20, 182n97, 198n78
empire. *See* Roman Empire
encyclical. *See* Laudato Si'
environmental: backdrop, 4, 46; ethics, 16, 41–42, 44–51, 52, 57, 136, 167n47; policymaking, 15; racism, 14
Ephesians, 197n72
Erikson, Kai, 61, 67, 173n45
eschatology, 9, 56–57, 72, 74, 82, 89, 106–7, 120, 123–24, 155n19, 175n2, 182n98, 190n82, 190n87, 194–95n46. *See also* lament (eschatological), trauma (eschatology and)
ESST. *See* Earth systems stress trauma
eternity, 107, 121–24, 195n46, 196n63
ethics, 16, 18, 126, 131, 136–37, 141, 203n39,

203n43; banality of, 48–49, 166n43; rationality of, 48–51, 52, 166n44. *See also* Anthropocene (ethics in the), climate change (ethics of), environmental (ethics)
eucharist, 18, 138–40, 178n41, 183n110, 188n59, 205n60
Eve, 119, 152n66, 197n73
evil, 49–50, 124; anthropogenic, 47, 50, 147n19; banality of, 166n43; moral, 44–47, 165n36; natural, 45–47, 50, 164n22, 165n36. *See also* incarnation (in evil)
evolution, 4, 12, 27, 45, 82, 85–86, 97
Exodus, 182n93
extinction, 1, 3–4, 6, 19, 41, 44–45, 48–49, 57, 64, 75, 77, 96, 113, 125, 133, 135, 138–39, 164n19, 167n47, 187n32. *See also* mass extinction
Extinction Rebellion, 42
Eyerman, Ron, 65
Ezekiel, 70, 154n5, 174n66

fall, 45, 74, 86, 164n22, 165n31, 170n3, 194n38
Farley, Wendy, 49–50
fatalism, 17, 74, 90, 133–34, 141, 176n4
feedback loops, 6, 28, 46, 112
feminism. *See* ecofeminism, theology (feminist)
Feuerbach, Ludwig, 36–37, 161n109
Fiddes, Paul, 200n107
flashbacks, 5, 84, 94–95, 111–112, 192n10, 194n37
flesh: of the Earth, 17, 98, 124–26, 136, 139; ecological, 94, 96–97, 99, 102, 106, 108; Gregersen on, 11–12, 97–98, 186n25; human, 90, 94, 103, 186n25, 188n54; matter as, 6, 96–97, 102, 108; rupture of, 4–7, 11, 16–17, 64, 75, 92, 93–109, 130, 148n24; sharing in the, 98–99; versus bodies, 94–96; of the world, 100, 107. *See also* Christ (flesh of), witnessing (enfleshed)
forest fires, 69, 95. *See also* Australian wildfires
fossil fuels, 106, 114, 132, 137
Foster, John, 50
Francis. *See* Pope Francis, Saint Francis
Fretheim, Terence, 21
Fukushima Texture Pack, 93–94, 108
future: generations, 125; imagination of the, 2, 5, 111, 125, 135, 137, 191n4; violence, 50, 77, 136–37, 192n10, 200n101

Gaia theory, 15, 25, 28–29, 31, 35–36, 39, 158n52, 184n10
Genesis, 32, 159n75
geotrauma, Land on, 147n21

glaciers, 6, 79. *See also* Okjökull
global warming, 42–43, 112, 131, 192n14
God: as companion, 87–88; as co-sufferer, 88–91; protest against, 70, 71; Word of, 119, 190n86, 197n73, 197n75. *See also* body (of God), trauma (of God), violence (done to God)
godforsakenness, 72, 84–85
Golgotha, 56
Good Friday, 79, 115
gospel: of Matthew, 72, 81–82, 84, 145n1; of Luke, 21, 122, 201n109; of John, xi, 12, 77, 97, 121–22, 129, 179n47, 185n21, 186n25, 199n96
Gould, Stephen Jay, 145n3
Graham, Elaine, 8, 150n47
greenhouse gases, 3, 6–7, 67, 112, 131, 133
Greenland ice sheet, 6, 112
Gregersen, Niels, 11–12, 17, 88–90, 97–99, 106–7, 116–118, 121, 126, 151n59, 152nn65–66, 185n20, 186n25, 187n32, 190n82, 190n86, 191n97. *See also* flesh (Gregersen on)
Gregory of Nazianzus, 105, 189n81
Gregory of Nyssa, 90
Gregory the Great, 90, 183n105
grief. *See* ecological (grief)
groaning. *See* creation (groaning of)

Habel, Norman, 72, 103–4
Haitian earthquake, 165n36
Hamilton, Clive, 41–42, 48–49, 113
Haraway, Donna, 74–75, 77, 82, 90, 92, 134, 175n2, 176n4, 176n9
Harding, Stephan, 28, 157n51
Harris, Melanie, 14
Harvey, Graham, 15, 30
Hatley, James, 203n43
Hayes, Katherine, 20–21, 37
Hayhoe, Katharine, 42
Hebrew Bible, 16, 20, 33, 36, 39, 70–72
Hebrews, 189n77
hell, 115, 124, 178n37
Herman, Judith, 7, 111, 113,
hermeneutics. *See* ecological (hermeneutics), trauma (hermeneutics)
Heyse-Moore, Louis, 28, 95, 184n10
Hill, Preston, 79–80
Hine, Dougald, 132
Holy Saturday, 79, 115–116, 177n30, 178n37
Holy Spirit. *See* Spirit
hope, 56, 74–75, 90, 92, 106, 124, 131, 134–36, 176n4, 203n34; against hope, 135; authentic, 135; Christian, 12, 89, 136; in the dark, 135, 203n29; false, 115, 134; imagination and, 18,

135; melancholic, 135; radical, 18, 135–37, 141; in ruins, 135; tragic, 135. *See also* hopelessness, Rambo (on hope)
hopelessness, 89, 124; De La Torre on, 134
Hosea, 19–20, 36, 174n66
Hull, John, 124, 199n96
human. *See* accompaniment (human), body (the human), flesh (human), psychology (human), soul (human), trauma (human)
hurricane: 192n14; Betsy, 66, 112; Ida, 66, 112; Isaac, 66, 112; Katrina, 16, 65–69, 73, 112, 172n36. *See also* Rambo (on Hurricane Katrina)
Hutton, James, 110

ice: ages, 113, 193n25; caps, 1, 29, 39; cores, 78, 114; sheets, 6, 96, 112–114. *See also* Antarctic ice sheets, Greenland ice sheet
I-Ens relationships, 36
I-It relationships, 35
imago Dei, 103
imago mundi, 103
immanence, divine, 106–7, 185n20
impassibility, 31–32, 54–55, 91, 168n77, 200n107
incarnation: and chronocentrism, 116–118; and the creeds, 185n22; in evil, 106; in matter, 11, 97–99, 102; promiscuous, 104; as salvific, 105–6, 152–53n68, 182n99; supralapsarianism and, 12, 152n66. *See also* annunciation-incarnation event, deep incarnation, pancarnation, witnessing (enfleshed)
Indigenous communities, 23, 31, 33, 35, 45, 155n14, 160n99
intrinsic value, 27, 87, 157n35, 170n3
IPCC, 3, 7, 192n14
Irenaeus, 119–20, 181n75, 197nn73–75, 198n81
Irvine, Richard, 114
Isaiah, 21, 71, 154n15, 155n11, 175n67, 182n93
I-Thou relationships, 35–36

Jamieson, Dale, 166n44
Jenson, Robert, 197n70
Jeremiah, 154nn3–5, 174n66, 182n93
Jesus: historical, 81, 118; as witness, 81–83; of Nazareth, 117, 119, 121. *See also* Christ
Job, 23, 156n24
Joel, 154n5
Joerstad, Mari, 33–34, 36–37, 162n116
Johannine prologue, 26, 94, 96–97, 185nn19–20, 186n25
John. *See* gospel (of John)
Johnson, Elizabeth, 17, 81–82, 84, 88–89, 97–99, 161n106, 182n93, 186n27, 188n62

Jones, Serene, 9, 16, 47, 59–60, 62, 64, 68–69, 71, 73, 83–84, 100, 134, 149n41, 150n43, 151n58, 170n3, 173n47, 175n2, 183n103, 197n71
Jung, Carl, 158n63
Justin Martyr, 181n75, 197n75

kairos, 196n63
Kansteiner, Wulf, 63
Kaplan, Ann, 2
Katangole, Emmanuel, 135
Katrina. *See* hurricane (Katrina)
Keller, Catherine, 16, 41, 51–53, 56–57, 104–5, 107, 118–20, 134, 168n67, 176n9, 178n42, 196n68, 197n71, 201n108
Keshgegian, Flora, 118, 192n12, 194n37, 196n63
Kidd, Erin, 187n50
Kingsnorth, Paul, 132
Kwok Pui-lan, 13

lament: of the land, 16, 19–20, 22; eschatological, 56, 124–25; psalms of, 70–71, 139
land. *See* cry (of the land), lament (of the land), trauma (of the land)
Land, Nick. *See* geotrauma (Land on)
landscape. *See* scars (on the landscape), trauma (of the landscape)
Lange, Dirk, 139, 150n43, 203n39, 204n53, 205n59
language: anthropomorphic, 22, 24–25, 31, 173n47; of crisis, 42, 163n7; rupture of, 5, 75–76, 100, 148n24. *See also* trauma (language of)
Latour, Bruno, 24–25, 46, 156n30, 204n49
Laub, Dori, 5–6, 17, 75, 89, 138, 148n27, 183n100
Laudato Si', 16, 23–24, 36, 39, 44, 56–57, 156n27, 156n29, 158n66
Lear, Jonathan, 135, 137, 203n34
Leidenhag, Joanna, 15, 27
liberation. *See* theology (liberation)
Linzey, Andrew, 186n25
liturgy, 18, 118, 138–40, 203n39, 204n53. *See also* memory (and liturgy)
Logos, 12, 26, 31, 39, 96–97, 118, 121–22, 159n76, 178n37, 185n20; rationality of the, 25–26, 136
Logos-sarx, 105
Lopresti-Goodman, Stacy, 61
Lovelock, James, 15, 28–29, 33, 35
Luke. *See* gospel (of Luke)
Luther, Martin, 123, 187n32, 188n59
lynching, 14, 104, 175n71

macrocosm. *See* microcosm/macrocosm
Macy, Joanna, 30
Margulis, Lynn, 28
martyrdom, 76, 79–80, 140, 177n30, 177n33
Mary, mother of Jesus, 119, 197n73
mass extinction, 3, 19, 48, 64, 75, 113, 133, 135, 138. *See also* extinction
Mass on the World, 140
Mattes, Eva and Franco, 93
Matthew. *See* gospel (of Matthew)
Maximus the Confessor, 90
Mbembe, Achille, 15
McCarroll, Pamela, 16, 51–52, 57, 135
McDaniel, Jay, 88
McFague, Sallie, 11, 53–55, 164n25, 169n83, 183n111. *See also* metaphor (McFague on)
McPhee, John, 191n4
memory: divine, 107–8, 191n92; and erasure, 124–5; and liturgy, 138–39; multidirectional, 64; somatic, 93, 95, 108, 184n10. *See also* Earth (memory of the), trauma (memory of)
metaphor: anthropomorphic, 22–23, 25, 36–39, 161n108; Gaia and, 29; Latour on, 24; McFague on, 53, 169n83; of voice, 22, 36; of the wake, 14. *See also* trauma (natural metaphors for)
methodology, of trauma theology, 1–2, 7–10, 11–12, 42, 46–47, 73, 86, 129, 135, 150n43
microcosm/macrocosm, 17, 26, 90–91, 102, 183n107, 188n60
Midgley, Mary, 29
Minns, Denis, 197n75, 198n78
mirroring, Christology of, 83–84, 100, 180n65, 183n103, 197n71
Moltmann, Jürgen, 188n56, 188n63, 190n87, 200n107
Morton, Timothy, 43
Moss, Candida, 199n96
Mother Earth, 23–24

Nagel, Thomas, 157nn42–43
Narine, Anil, 110, 125
Nash, James, 103, 188n60
natural disasters. *See* disasters (natural)
nature: anthropomorphisms of, 21, 38, 78; silencing of, 21, 31, 37, 39, 138, 155n18, 201n109; speaking for, 138; Spirit in, 31. *See also* disvalue (in nature), nature/culture binary, soul (in nature)
nature/culture binary, 46–48, 50, 66–67, 73, 130, 147n11, 173n44
Neoplatonism, 25
New Orleans, 65–66, 112, 172n36

New Testament, 11, 21–22, 97, 185n20, 194n38
Nicaea, council of, 185n22
Nixon, Rob, 112, 195n53
nonhuman: agency, 21, 27, 31–34, 39, 46, 156n30, 165n36; agnosticism about subjectivity of the, 15, 34–36, 39; animacy of the, 31–32, 34–36, 38, 159n74, 161n101; creation, 21, 84, 98, 155n16, 170n13; voice of the, 21, 24, 27, 33, 161n110; witnessing, 137, 193n25. *See also* cry (of the nonhuman)
nonidentical repetition, 119–20, 139
Northcott, Michael, 44–45, 159n76
nuclear: disaster, 93; weapons, 114

O'Donnell, Karen 4–5, 7–8, 62, 64, 69, 75, 138–40, 148n24
oil spill. *See* Deepwater Horizon oil spill
Okjökull, 139
Oliver, Kelly, 77–78, 101, 137
Orange, Donna, 50, 173n42
Origen, 25
Orthodox. *See* theology (Eastern Orthodox)
orthopraxis, preceding orthodoxy, 8, 24

Pachamama, 23
paganism, 26, 32, 118
Page, Ruth, 17, 87–88, 91, 181nn85–87, 181n90
pancarnation, 17, 94, 104–8, 190n82
panentheism, 54
panpsychism, 15, 24, 25, 27–28, 35, 39, 157n42
pantheism, 24, 31–32, 159n71
parables, 82, 179n56, 181n82
Paris Agreement, 3, 132
Parliament of Things, 204n49
passibility. *See* impassibility
pastoral. *See* theology (pastoral), trauma (pastoral response to)
pathetic fallacy, 37
Patriarch Bartholomew, 164n25
patristics. *See* theology (patristic)
Paul, Saint, 21, 117, 155n16, 156n27, 185n20, 186n23, 195n56
Peacocke, Arthur, 88
personification. *See* anthropomorphism
Philippians, 185n22
Pihkala, Panu, 2, 135, 167n53
Pilate, Pontius, 179n47
Plato, 25–27, 197n75
pneumatology, 10, 31, 54–55, 79–80, 159n77, 178n42, 179n44. *See also* ecological (pneumatology), Spirit
polytheism, 33, 159n71
poor. *See* cry (of the poor)

Pope Francis, 16, 23–24, 35, 40, 44, 56–57, 60
postcolonial traumatic stress disorder, 14, 153n78
post-traumatic growth, 69, 112
post-traumatic slavery syndrome, 14, 153n78
post-traumatic stress disorder, 2, 61, 65, 67
Pound, Marcus, 150n43
practical theology. See theology (practical)
pre-traumatic stress syndrome, 2
PreTSS. See pre-traumatic stress syndrome
problem of evil. See evil
process theology. See theology (process)
progress narratives, 116, 194n40
prophets, in scripture, 19–20, 21, 71
Psalms, 21, 70–71, 139, 155n11, 175n70. See also lament (psalms of)
psychology: and anthropomorphism, 38–39; of Christ, 100; human, 19, 161n109; Jungian, 158n63. See also trauma (psychology and)
PTSD. See post-traumatic stress disorder

racism. See environmental (racism), trauma (racism and)
Radford Ruether, Rosemary, 102
radical hope. See hope (radical)
Rae, Eleanor, 169n88
Rahner, Karl, 105, 200n107
Rambo, Shelly: on Calvin, 123; on Doubting Thomas, x–xi, 126; on the expansion of trauma theology, 62–64, 172n22; on hope, 134–35, 137; on Hurricane Katrina, 16, 65–68, 73; on soteriology, 10, 131, 151n54; on temporal trauma, 5, 114–116, 118, 126; on theodicy, 50; on theology from the middle, 48, 56; on three ruptures, 4, 75, 148n24; on witnessing, 17, 76–77, 79–81, 100, 122, 138, 177n17
recapitulation, 16–17, 111, 118–20, 126, 197nn71–75
redemption. See salvation
Reid, Duncan, 151n59, 188n55
relational anthropomorphism. See anthropomorphism
relationship. See I-Ens relationships, I-It relationships, I-Thou relationships
remaining. See suffering (remaining with), witnessing (remaining as)
repetition, 119–20; of liturgy, 18, 118, 138–40, 203n39, 204n53. See also trauma (repetition as symptom of)
representation, 90–91, 102–4, 106–7; by Christ, 17, 92, 94, 101–3, 108, 188n55; and identity, 102; political, 204n49; of the unrepresentable, 99. See also Earth (representation of the)
Resurrecting Wounds, 80
resurrection: for all creation, 89; ambiguity of, 10, 53, 131, 134, 151n54; appearances, 100, 137; ongoing, 114–118; wounds, xi, 80, 111, 122–26, 132
Revelation, 124, 168n67
Richardson, Michael, 137, 146n4, 147n20, 193n25
Rivera, Mayra, 185n19, 186n25
rock record, 7, 17, 111, 113, 125, 193n25, 201n109. See also consciousness (of rocks)
Rolston, Holmes, 17, 85–86, 91, 119, 133, 181n75, 181n83, 197n75
Roman Empire, 13
Romans, 21, 155n16, 156n27, 203n35
Roszak, Theodore, 15, 29–30, 33, 35, 158n66, 160n99
Rothberg, Michael, 64, 171n13, 172n28
Rubenstein, Mary-Jane, 32, 35, 166n46
rupture. See communication (rupture of), flesh (rupture of), language (rupture of), Rambo (on three ruptures), time (rupture of)

sacrament, 139–40, 190n84, 205n61
sacrifice, 76, 86, 138
Saint Francis, 23–24
salvation: cosmic, 12, 85; co-suffering and, 88–89; healing and, 189n78; history, 115, 118; linear models of, 10, 66, 115, 120, 194nn38–39; reconfiguration of, 10, 12, 53, 58, 86, 88, 92, 101–2, 108, 131. See also incarnation (as salvific), Rambo (on soteriology)
Santmire, Paul, 35
sarx. See flesh
scars: Anthropocene, 17, 110, 124–26, 137; on Christ's body, 80, 123; on the landscape, 6, 79, 96; wounds versus, 123, 199n95
Scarsella, Hilary, 71
Schneider, Laurel, 104, 188n55
scripture. See Hebrew Bible, New Testament, prophets (in scripture)
secondary trauma. See trauma (secondary)
Sermon on the Mount, 81
sexual and domestic abuse, 1, 5, 14, 53, 76, 94, 152n67, 170n3. See also trauma (sexual abuse and)
Sharpe, Christina, 14, 153n76
sin, 9, 12, 20, 70, 97, 102, 105, 123, 152n66, 165n31, 169n87, 182n99, 186n25, 187n32. See also ecological (sin)
Skrimshire, Stefan, 112, 125–26, 136

slow violence, 112–113
Sobrino, Jon, 81, 180n67, 183n101
solidarity: of Christ, 13, 15, 71, 83–84, 86, 91, 98–99, 101, 108, 188n61; divine, 54, 84, 88, 94, 103, 106, 178n37, 182n99; planetary, 175–76n3
Sollereder, Bethany, 51
Solnit, Rebecca, 133, 135, 203n29
Sorrell, Roger, 24–25
soteriology. See salvation
soul: human, 25–26, 105; in nature, 25–26, 32, 34–35. See also, world soul
Southgate, Christopher, 17, 49, 88–89, 106–7, 164n19, 165n31, 167n47, 181n75, 182n98
Spirit, 26–27, 31–32, 39, 79, 98, 107; disembodiment of the, 79–80, 178n37, 178n43; embodiment of the, 54. See also ecological (pneumatology), nature (Spirit in), pneumatology, wounding (of the Spirit)
Spirit and Trauma, 65, 67, 76, 80, 135
Stanley, Elizabeth, 52
Stations of the Forests, 180n69, 204n56
stewardship, 11, 52, 82
Stoic philosophy, 26, 185n20
subjectivity: constructed through witnessing, 78; and panpsychism, 27. See also Earth (subjectivity of the), nonhuman (subjectivity of the)
suffering: acknowledgment of, 9, 17, 53, 56, 58, 123; glorifying, 12, 87, 115, 131, 138; incomprehensibility of, 42, 50, 57; ongoing, 1, 48, 134, 197n71; remaining with, 10, 17, 53, 75, 80, 89, 92, 137, 141. See also co-suffering, Earth (suffering of the), salvation (co-suffering and)
Szerszynski, Bronislaw, 193n25

Tanner, Kathryn, 152n68, 182n99
Tansey, Mark, ix–xi, 129, 145n3
Taylor, Bron, 156n35
Teilhard de Chardin, 140
temporality. See time
tenebrae creationis, 98
Tertullian, 105
testimony, 76, 78, 83–85, 89, 100–1, 140, 177n17, 183n102
theodicy, 9, 16; anti–, 49–50; evolutionary, 89, 182n98, 187n32; natural, 42, 44–51, 52, 57, 167n47. See also Rambo (on theodicy)
Theokritoff, Elizabeth, 139
theology: constructive, 7–9, 149n40, 150nn43–44; disability, 96, 124, 199n96; Eastern Orthodox, 139; ecological, 41–58,

73, 132; feminist, 12, 76, 153n68; liberation, 23–24, 81, 182n99; pastoral, 49, 51, 123; patristic, 90, 105, 122, 188n60, 191n97; practical, 7–8, 51, 131, 150n44; process, 119; systematic, 8–11, 182n99; womanist, 12, 76, 153n68. See also methodology (of trauma theology), Rambo (on the expansion of trauma theology), Rambo (on theology from the middle)
Thomas. See Doubting Thomas, Rambo (on Doubting Thomas)
Thompson, Deanna, 9, 62, 64, 170n4, 171n5
time: cyclical, 118–20, 139, 197n70; deep, 17, 48, 110–111, 121–26, 191n4, 193n25, 198n85; helical, 17, 111, 118–21, 126, 197nn70–71; imagination and, 125–26, 137; linear, 5, 17, 111–20, 192n12; rupture of, 4–7, 110–27; spiral of, 118–119, 126. See also trauma (temporal)
Tinker, George "Tink," 155n14
Tonstad, Sigve, 21
tragedy, 1, 49–50, 79, 89, 135, 141, 167n53, 181n83
trauma: afterlife of, 116, 151n54; in chimpanzees, 2, 61–62; climate, 4, 43, 173n42; collective, 2–3, 51, 61–67, 130, 140, 171n8; cultural, 2, 61–67, 171n8; Earth, 2, 19–40, 51, 168n67; in elephants, 2, 61–62; environmental, 4, 55, 59, 73; eschatology and, 56–58, 169n91, 170n95; experience of, 5, 8–9, 13, 19, 26–27, 30, 62, 94–95, 111, 138, 149n41, 151n58; fracture entailed by, 10, 83–84, 100, 129, 192n12; fragmentation entailed by, 5, 9, 26, 30, 70, 111, 120, 140; of God, 17, 53–54, 91; hermeneutics, 70–71; human, 2–6, 26, 69–71, 75, 83, 94–95, 111; imagination in response to, 9, 135; inaccessibility of, 5, 68, 136, 203n39; inarticulability of, 10, 49, 52, 130; interconnection of, 15, 60, 65–68, 73; irreparability of, 41, 52–53, 55–57, 130; lack of teleology in, 4, 21, 86–87, 120, 156n27; of the land, 70; of the landscape, 2, 66–67, 69, 73, 93–94, 96; language of, 16, 42–44, 52, 57, 164n18, 168n67; latency of, 7, 111–112; marks of, 6, 56, 66–67, 100, 104, 107–8, 116, 122–24, 180n64, 191n98, 199n96; memory of, 5, 69, 89, 95, 111, 194n37; multispecies, 62, 65; narrative of, 5, 138, 148n28; natural metaphors for, 69, 72, 93–94; ongoing, 9, 62, 93–94, 114, 120; pastoral response to, 7, 149n40; permanence of, 7, 17, 56–58, 111–114, 116, 118, 121–26, 139, 141; planetary, 4, 28, 53, 78, 125; psychology and, 5–6, 26, 29, 59,

61–62, 95; racism and, 14–15, 62, 66, 170n3; recurrence of, 5, 17, 111–112, 114–116, 118–20, 126–27, 130, 139, 192n10; remembrance of, 5, 56, 61, 113, 123–24, 139, 171n8; repetition as symptom of, 70, 111–112, 118; science of, 62–63; secondary, 62; sexual abuse and, 5, 53, 124; sharing in, 17, 98–99, 100–2, 107, 130, 190n86; societal, 62, 73; survivors of, 10, 13, 83, 87, 101, 103, 106; symptoms of, 2, 5, 95, 113, 115, 117; temporal, 17, 56, 121–22, 126, 194n37; unassimilable, 41, 52–53, 77; unthinkability of, 166n46; vicarious, 80. *See also* Christ (as subject of ecological trauma), climate change (as traumatic), Earth systems stress trauma, ecotrauma (Keller on), geotrauma (Land on), methodology (of trauma theology), postcolonial traumatic stress disorder, post–traumatic growth, post–traumatic slavery syndrome, post–traumatic stress disorder, pre–traumatic stress syndrome, *Spirit and Trauma*, *Trauma and Grace*
Trauma and Grace, 9, 59, 83
Triduum, 62, 79, 115, 139
Trinity, 26, 55, 121, 157n40, 179n44, 200n107
trouble, staying with the, 74–75, 77–78, 82, 122, 134, 175n2, 176n4, 176n9
tsunamis, 69, 93
Tumminio Hansen, Danielle, 16, 51, 53–55, 57–58, 91, 124
Tylor, Edward Burnett, 33

Uehlinger, Christoph, 20–21
unassumed is the unhealed, 105–6, 189n81
United Nations Secretary General, 42
Urbaniak, Jacob, 179n50

value. *See* disvalue, intrinsic value
van der Kolk, Bessel, 5–6, 95, 101, 148n31
Verlie, Blanche, 78, 133
Vibrant Matter, 33
violence: of the cross, 12–13, 83, 87; done to God, 54; unforgettable, 56, 124. *See also* future (violence), slow violence
Vogt, Markus, 48
voice. *See* Earth, (voice of the), metaphor (of voice), nonhuman (voice of the), *The Voice of the Earth*
The Voice of the Earth, 29
von Speyr, Adrienne, 79, 115

Wainwright, Elaine, 72, 84
Walking Buffalo, 33
Wallace, Mark, 16, 31, 34–35, 51, 54–58, 91, 104, 158n52, 159n75, 159n77, 160n88, 160n101, 169n83, 169n87, 178nn42–43
Ward, Graham, 178n41, 183n110
Warner, Megan, 70
Westhelle, Vítor, 195n48
White, Benjamin, 2
White, Bob, 165n36
White, Lynn 32
Widdicombe, Peter, 200n107
Wiesel, Elie, 103
wildfires. *See* Australian wildfires
Wiles, Maurice, 189n78
Williams, Delores, 152n67,
witnessing: activity of, 17, 76, 90–91, 137, 180n65, 183n103; collapse of, 5–6, 17, 75; of the disciples, 79–80, 100, 122; duration of, 92, 122, 127, 134; embodiment and, 79–80, 83–85, 100, 138, 178n37; enfleshed, 93–109; imitation as, 17, 76–77, 79–80, 92, 93, 100, 138, 140, 177n17, 177n30; material, 93–94; ongoing, 122–23, 125, 195n46; proclamation as, 17, 76–77, 89, 92, 93, 100, 138, 140, 177n17; remaining as, 12, 18, 66, 77–79, 81, 84, 108, 115, 122–24, 136–37, 176n9; science as, 204n49; silent, 37, 81, 93, 180n67. *See also* bearing witness, *Bearing Witness*, Christic witnessing, cruciform (witnessing), Jesus (as witness), nonhuman (witnessing), Rambo (on witnessing), subjectivity (constructed through witnessing)
womanism. *See* theology (womanist)
Woodbury, Zhiwa, 43
Woolbright, Lauren, 39, 67
world soul, 15, 25–27, 30–32, 39, 197n75
wounding: glorifying, 123–24; incorporation of, 17, 93–94, 101, 107–9, 130; inscription of, 101, 107–8, 124, 191n92; of the Spirit, 54–55, 91, 169n87; structural, 61, 63. *See also* Christ (wounds of), *Resurrecting Wounds*, resurrection (wounds), scars (wounds versus)
Wyman, Jason, 8

YHWH, 20–21
Yusoff, Kathryn, 148n21

Zechariah, 124
Zylinska, Joanna, 137, 204n45

TIMOTHY A. MIDDLETON is a Tutorial Fellow in Theology at Regent's Park College, University of Oxford. His research focuses on the intersections of theology and religion with science, nature, and the environment.

www.ingramcontent.com/pod-product-compliance
Lightning Source LLC
Chambersburg PA
CBHW031145020426
42333CB00013B/523